MEMOIRS OF GOLD-DIGGING IN AUSTRALIA

MEMOIRS OF GOLD-DIGGING IN AUSTRALIA

Seweryn Korzelinski

Translated and edited by Stanley Robe

Foreword and Notes by Lloyd Robson

University of Queensland Press

Typeset by Academy Press Pty Ltd, Brisbane
Printed and bound by Silex Enterprise & Printing Co., Hong Kong

Distributed in the United Kingdom, Europe, the Middle East, Africa, and the Caribbean by Prentice-Hall International, International Book Distributors Ltd, 66 Wood Lane End, Hemel Hempstead, Herts., England

National Library of Australia Cataloguing-in-Publication data

Korzelinski, Seweryn, 1802–1876.
Memoirs of gold-digging in Australia.

Index
ISBN 0 7022 1346 2

1. Gold mines and mining – Victoria – History.
2. Gold miners – Victoria – Biography. 3. Victoria – History – 1852–1856. I. Robe, Stanley, 1908–, tr.
II. Title.

994'.5'031'0924

Contents

List of Illustrations

Foreword

This translation of Seweryn Korzelinski's story of his adventures on the Victorian goldfields is the result of quite an exciting find—a first-hand and close account of what it was like to be a gold miner and a Polish immigrant in the early 1850s.

Stan Robe's rendering is of considerable historical interest, not least in terms of Korzelinski's detailed report on the Red Ribbon movement at Bendigo and in the author's reflections on the nature of unrest and rebellion on the goldfields. Especially memorable is Korzelinski's reaction to the move to enlist him in the rebellion.

Of further value is the author's description of the Aboriginals and his encounters with them. Indeed, he is always interesting and entertaining in his alertness and curiosity, whether it is reaction to the Aboriginals' way of life, to the primitive settlement called Melbourne, to his fellow immigrants or to the quality of the Australian bush.

Korzelinski must have been the newest of new chums ever, and one of the most delightful aspects of this book is the gradual transformation of the author from officer and gentleman to resourceful digger.

Korzelinski's work is important in other areas, I think. The first is his intriguing dislike of the English. Does this lead him to overemphasize the importance of immigrants other than English? Perhaps, but it is most salutary to have an account of this period which draws attention to the extraordinarily cosmopolitan character of the population of the goldfields.

The second area of significance emerges in Korzelinski's compelling account of the day-to-day life of the miner. Here the reader finds out exactly how gold was searched for, how a shaft was dug, what it was like to live in a perpetual state of camping, what men ate and did on the rare occasions when they lifted their eyes from searching for gold. The magic of, and compulsive search for, the precious metal has seldom been described so well.

Thirdly, for those who have been led to believe that life on the goldfields was surprisingly law-abiding and "English", Korzelinski has some surprises. It is quite striking how frequently he feels it necessary to draw attention to incidents involving violence and, indeed, murder.

The dangers of travelling alone in the bush are well evoked and some splendid vignettes are memorable. Who else but an old soldier would know how close a bullet had come to him by the slightly different noise it made? Who else but a non-English immigrant would be interested to describe the different sorts of food eaten by the national groups or the different sorts of songs they sang and tales they told?

The veteran Korzelinski reveals his own preoccupations and assumptions with engaging clarity. He is a survivor of the 1848 revolutions in Europe and the fact that his heart is ever with his beloved Poland clarifies and sharpens his perception of Australia.

The man who had rendered Korzelinski's work into English for the first time ever is also of Polish origin but probably exceeds even his fellow in energy. He certainly does in variety of interests.

Stan Robe immigrated to Australia seventy years after Korzelinski and has been impressed by how similar Australia is now to the Victoria of the 1850s. Stan sees Australia as the same sort of melting-pot it was then. He has become convinced that Australians should be more aware of how important it is to research further into the origins, expectations and experiences of the vast number of ethnic groups which now exist in Australia.

We must get to know our history in order that we may more fully understand how we came to be what we are, Stan Robe observes. He considers that the views and opinions of the multifarious groups in the nation may be linked through a perception of patriotism.

The life of Stan Robe in Australia began in 1926. From a family which has a tradition of involving itself in progressive movements in Russia and then in Poland, he inherits a deep awareness of social justice and injustice.

He worked in the textile industry in Melbourne and improved himself by night study at technical school. In the Carlton of the 1930s the Depression was everywhere apparent and Stan could see how it severely limited human freedom.

As a humanist, he found six years in the A.I.F. less than inspiring in some ways. He joined the Fellowship of Australian Writers by correspondence at the end of his time in the army and was a founding member of the Australasian Book Society and the Realist Writers group.

After the war he was active in helping get pensions, by way of restitution, for first Jewish and then other victims of the Nazi concentration camps who settled in Australia. He was the first organizer of the Hakoah Soccer Club (the first ethnic sports club in Victoria) and secretary and delegate to the Soccer Association.

It is characteristic of Stan Robe that having earlier been responsible for new inventions in the textile industry, expecially in the knitting sector, and acting as a consultant in that industry, on his retirement he turned to a study of art, astronomy and physics—and this translation of Korzelinski's memoirs.

Stan Robe sees this book as the first of several concerned with historical research and translations. Korzelinski did not stay in Australia but fortunately Stan Robe did.

(DR) L.L. ROBSON
History Department,
University of Melbourne

Acknowledgments

I would like to express my thanks and appreciation for help and encouragement in putting this book together to John Iremonger, Lech Paszkowski and Lloyd Robson.

I would also like to thank the executors of the La Trobe Library Collection, State Library of Victoria for their kind permission to reproduce the illustration in this book.

Translator's Introduction

Seweryn Korzelinski was a Pole, a Polish patriot. To understand him let us take a brief glance at Polish history. Poland had been an independent nation in eastern central Europe since approximately 900 A.D., with a checkered history. At times one of the mightiest nations in Europe, the protector and succour of Austria against the Turks who encircled Vienna; and the recipient of a delegation of Russian nobles asking to place a Polish prince on the Russian throne. At other times subject to pressure of both an internal and external kind.

In the year 1772 the first of three partitions took place, in the form of annexation of Polish territory by Prussia, Austria and Russia, taking nearly 35 per cent of the population from the Polish State. This partition was followed by two more in 1793 and in 1795. Partition and annexation were accompanied by attempts by force to assimilate the Polish population into the occupying power's political and ethnic structure, attempts bitterly resented by the Poles. There were armed insurrections against the occupants, stirred by the patriotism of the Poles and no doubt helped by the winds of revolution and liberalism from France.

Seweryn Korzelinski took part in such armed actions. Born in 1804, he gave up the prospect of a legal career to take part in an insurrection in 1830 and then in the Hungarian insurrection in 1848, where a Polish division helped the Hungarians against Austria. He attained the rank of a major and was interned in Turkey when the Hungarian campaign ended unsuccessfully. Eventually he managed to make his way to England. Tramping through Europe from Poland to Hungary then to Turkey to end up in England, meeting various people in those countries, appreciating that people are different yet all brothers in arms or brothers in need, was a good training for mingling with the cosmopolitan crowd on the gold-diggings in Victoria later on.

England in those days was no paradise, and it was particularly difficult for migrants without a knowledge of English and without a trade. Korzelinski and a group of other Polish veterans of the Polish and Hungarian campaign decided to migrate to Australia after hearing about the opening of the goldfields in Victoria. He embarked at Liverpool on 21 August 1852 on the *Great Britain* with eleven companions on the way to Australia. He left Australia in 1856 to return to his homeland after being granted an amnesty by the Austrians, and died in Poland in 1876.

His memoirs, written from diaries kept while in Australia, were first published in Cracow in 1858 in two volumes. The work, meant for reading by Poles, of necessity used Polish phonetic spelling and pronunciation of English names. It is filled with comparisons and personal reflections on life in Poland and Europe. As an old insurrectionist, the author was well-trained in conspiracy: he rarely used the full names of his compatriots in order not to cause hardship to their families. The same habit caused him to often use only the first letter of some localities in his diaries, particularly when describing political unrest or turmoil in Australia. This habit is responsible for a confusion in his memoirs in the names of Ballarat and Bendigo when describing the miners' protest action against the government of the Colony of Victoria.

The book is naturally angled at the Polish reader and consciously intended to describe the strange, exotic conditions in the distant, hardly-known land of Australia. In spite of this, or because of it, it gives a vivid "inside" picture of the goldfields and Australia generally in the mid-fifties of the nineteenth century. Not a cold picture of official reports, bureaucratic pedantry and sentimental or learned newspaper articles, but a warm, humane description of the vivid, turbulent life of a colony rapidly growing and expanding; of the sweat and toil; of the misery and of good fortune. Indeed, some nearly prophetic reflections of the writer on the future of Australia are well worth looking into.

So much about Seweryn Korzelinski, a Polish pioneer, if only of short duration, in Australia. Korzelinski left Australia in 1856 as a mature man. I arrived in Australia in 1926 as a young boy—seventy years later. It was another fifty years before I got to read the book about the Australia of 120 years before. Its impact has been great enough to make me render the part of the memoirs dealing with life in Australia into English.

There were good reasons for my attempt. One, already mentioned, was the rich yet down-to-earth description of life in those days on the goldfields, the historical material which will I hope enrich our knowledge of those days. But there were also other reasons. I became vividly aware that Australia, even in those early days, was a mixture of ethnic groups, each giving some share to what was to become the "Australian tradition", and I feel this fact should be made more widely known. History seems to repeat itself: we are again, 120 years later, a melting pot of ethnic groups, each capable of adding something of value to the growing and expanding Commonwealth, each group introducing new skills, new habits and new outlooks, but all of them looking upon this country as the homeland for their children and grandchildren. May we learn something from our past history to help organize this integration process a little better.

There are other points of interest in the book. Although the historian knows better, the man in the street is under the impression that the Australian tradition of democracy and mateship stems from the shearers and itinerant rural workers and their struggles and strikes against the squatters. Actually, as Korzelinski reveals, the idea and tradition were born in the sweat and toil of the goldfields. When the easily accessible alluvial gold gave out, the miners became the itinerant workers who carried the tradition further afield. Hammered out, or should I say dug out, on the goldfields, this tradition still lives. The very words "mate" and "digger" stem from the goldfields.

If I can speak for Korzelinski, having "lived" with him for over a year, day in and day out, while I was working on the book, I would like to dedicate this book in both our names to all the diggers and to all our mates.

STAN ROBE

1

I arrive in Melbourne

On the eleventh, very early in the morning, one can see on the horizon the dark shoreline of Australia. We approach rapidly, every hour details more clearly emerge from the watery haze. One can faintly discern hills called "Bald Head", and between two hills an empty space, probably an estuary of a river draining into the sea. All of it is too far away to judge the shape with any certainty or to name the apparition so like a *fata morgana.* We know however that this is the goal of our journey, that this is Australia.

Everyone on deck shows the joy the sight of land always brings after a long sea voyage. How much expectancy and uncertain hope shows in those faces around me. How many times will they become disappointed when stark reality replaces present uncertainty? Here then is the place where our fond dreams are to be fulfilled. How clever on the part of Providence to veil the future before a man's eyes, because many would perhaps prefer to jump overboard and so do away with themselves now, if they knew what was in store for them in the Eldorado of the antipodes, so highly praised in the European journals. But now one has sailed the vast ocean to reach here, it's not seemly to doubt, lose spirit and surrender to resignation. There is need to test the saying "He crossed the sea and perished near the shore." So let's go and see what next . . .

It's one o'clock in the afternoon. One can see medium-sized hills covered with virgin forest in the distance, a stream cataracts into the sea; there is not a village, nor even a dwelling in sight. The shores of this unknown country stretch ahead quietly and silently. Only imagination suggests hidden natives crouching amongst the trees, intently watching the smoke rising from the steamer's funnels.

The same view, the same sameness lasts for the rest of the day. About seven in the evening the environment changes—we are entering the approaches of Port Phillip Bay. On the left a lighthouse surrounded by trees, nearby a few houses and a high mast with a whole accoutrement of cross arms and ropes criss-crossing from the top of the mast to all the beams and furnished with balls of various sizes. By pulling on the string attached to the balls, one can move them up and down or sideways in any arrangement required, to send signals which can be read in Port Phillip or in Melbourne and which give advance notice of the arrival of shipping into the bay.

On the right the low sandy shore narrows into a sharp tongue. Here, as wide as a large river, flows and surges the sea and its white foam can be distinguished from the green waters. Near the end of the tongue lies a wrecked ship half-submerged in sand. Two men can be observed on the wreck busily collecting salvage. Before us the narrow and dangerous passage, reminder of which is the wreck. From one side a boat quickly approaches. It's the pilot. Just then the escaping steam surges up, the engine stops, the ship slows to a halt and the boat ties up. The agile pilot scrambles up like a cat and is on the deck in a wink; the captain passes the command over to him, because from the moment a pilot appears on deck, he takes sole responsibility for the safety of the ship.

The pilots always watch for signals from approaching ships at dangerous points. When it's foggy or stormy, when the ship is in danger of piling on the rocks or sand banks, those experienced sailors, acquainted with the locality and full of incomparable courage, man a boat be it day or night, stand up to the storm and other threatening dangers of the sea and guide the ships to safety. They do not discriminate as to the nationality of the ship in need of help. Many people owe their lives and their worldly goods to those brave men. Admittedly, every ship so led into port pays them a very large fee, but can there be a fee too great for saving one's life? We pass the narrows, find ourselves in a bay and all our fears are allayed.

Before us an unhappy view: on the right in a small port

are anchored three ships, two English and one American. Uselessly the crews stare towards the shore; none can disembark, shake the hand of a friend or embrace a relative awaiting their arrival . . . For there is disease aboard. Yellow fever daily kills those unfortunates who carried with them the germs of this terrible disease from Madras in West India. They have been in quarantine for several weeks and over two hundred have succumbed already. Poor devils. Their only hope is in God because human efforts are not sufficient. Everybody knows what epidemic means in such cramped quarters as a ship presents, where it is impossible to avoid contact and contact means death.

The bay gets wider the further we go. There to the left, about ten to twelve miles from Melbourne is the township of Gulong.[1] Directly in front of us is Phillip Tal[2] and Melbourne to the right. Night approaches and it's too dark to see anything except the flickering lights of ships at anchor. The *Great Britain* slows down and eventually stops opposite Phillip Tal.

The port is unlike anything I've ever seen so far. There is no sign of boats offering fruit, or boats with people curious to see the new arrivals. Possibly there is no fruit in Australia and as for the people, apparently they are too busy for idle excursions.

Some of the passengers prepare to go ashore to see what this promised land is like, but the price for the trip to the shore is so high that if similar prices were charged in London nobody would ever visit Greenwich or Cremorne Gardens. One could take an excursion on the Thames for two or three pence, but here the charge is five shillings.

News spreads about the exorbitant prices. Some say that to share a small mean room with two others one has to pay £2 per week. And so it seems that our group cannot afford to take excursions, nor live in rooms. But we don't need rooms for we have our tents, brought from London. It will be nice and romantic to live in a tent erected between evergreen trees in this permanent spring weather. The thought that there could be a storm, which might blow the tent down, or that rain might penetrate the tent doesn't enter our heads. We remember the times when during our army campaigns we slept under the stars. Now our large tent will be a royal shelter. We will have a fire near the tent and after gathering gold all day long, there we will rest in the evenings. What a poetic scene.

The main thing is to disembark and get to the goldfields. Alas, the captain is haggling with the barges for the cheapest price to transport passengers and their chattels to the shore, because it is the duty of the owners of *Great Britain* to land us. Hours pass and we still wait. The passengers who went to look at Melbourne are returning for the night and tell extraordinary tales. A pound of bread costs a shilling, an egg sixpence. Oysters, which in London cost four pence a dozen, here cost a shilling each.[3] A merchant in the city explained the situation to one of the passengers:

> You don't understand the situation here. In England a worker earns two shillings a day, here the wage is ten to twelve shillings. Therefore everything is at least six times dearer.

The news upsets us, because we have a very meagre cash supply. It seems that it will be difficult to buy bread and there will be no hope of having fried eggs. But why worry about high prices when one is healthy, strong and willing to work? Many a landed proprietor in Poland doesn't get more than ten shillings per day income from his land. We shall have gold so there is no need to worry.

Two of our countrymen, Leopold K. and Wladislaw K. visit us. They arrived in Australia a few months ago and they describe to us some of the details of life in Melbourne and on the diggings. They have been here only for a short period and do not have exhaustive information, but they were fortunate in the fact that they travelled to Australia on the same boat as the chief inspector of Port Phillip who was returning from England. Having good contacts in Melbourne, he helped them to get positions in the Mounted Police.

We were very happy to come across countrymen of ours, we did not expect such luck.

We could not disembark on the thirteenth because the

barge came only once from Melbourne. Everyone wanted to board it so only those who had little baggage left the ship. We remained until the next day, because our luggage was still in the holds. The barge was hired for the whole of the following day for a few hundred pounds so we were certain to land that day.

And so it came to pass. Our luggage was brought on deck early in the morning, trunks, cases, tents and all kinds of assorted gear. The passengers sorted out their belongings and put them together in heaps. As you can well imagine there was much noise and confusion at the embarkation, but it is a hundredfold worse at disembarking. Embarkation can last a week or longer as passengers arrive individually or in their groups and each arrival brings only his luggage. Here an attempt is made to disembark everyone in one day because feeding 650 passengers even a day longer represents a big expenditure.

A number of sailors then bring goods out of the hold, pulling everything up with the aid of ropes. As soon as a bundle or a case appears and the name of the owner is read out aloud, it is roughly thrown to the side and occasionally breaks open, strewing the contents on the deck. Someone standing some distance away suddenly hears his name read out, wants to get close to his property and starts pushing through. This provokes screaming, cursing and more confusion.

There! The barge arrives pulling two lighters with it to load those who have all their luggage assembled. We also load our chattels, although the cart and the pump are still below deck. We leave two of our companions to collect those and the rest of us board the barge. Four large bargefuls of luggage and passengers have not emptied the *Great Britain*. Only now can we assess the size and the capacity of our vessel. The captain and Mr Cook, first officer, stand at the steps farewelling the passengers.

We steam towards the shore, manoeuvring around the shoals marked by wooden posts, and turn right into the bed of the river Yarra Yarra. The river itself is narrow but deep and boats can sail right up into the city. The shores are low, nearly at water level, muddy and

The *S.S. Great Britain* on which Korzelinski sailed to Melbourne. This engraving was originally published in *Illustrated London News*, 15 July 1845.

Melbourne seen from the eastern side of Princes Bridge from the river Yarra in 1854. This lithograph by S.T. Gill was originally published in *Sketches in Victoria*, 1860.

overgrown with low bushes with a leaf similar to the cypress. On a shallow near the shore are pelicans, small water birds and two black swans. The bushes and the bird life are so different that by them alone we can tell that we are on the fifth part of the globe.

It takes an hour to travel into the town, the boats tie up and we carry our things ashore. We hire two carts costing ten shillings each and we go across a brick-faced bridge to the other side of the river where a number of tents belonging to passengers who arrived on other ships and who were as short of ready cash as we are, are pitched. It is already getting dark so there is no time to sort out our gear or to erect the tent. We have had nothing to eat all day so a fire is lit and tea made. With a piece of bread this makes our meal for the day.

The weather is beautiful and we are very tired; we shall sleep soundly under the trees. Everyone makes a makeshift bed and lies down, glad to rest on this new unknown soil. What will this new land, thousands of miles and many seas away from our motherland bring? None of us is a skilled tradesman, none of us is used to labouring, nor do we know the local habits or customs. What to do? Where to start? Those disturbing and unpleasant thoughts go through my mind, but eventually sleep chases the worries away.

2

This new country

It is hardly credible, when one looks at the growing population, rich commerce and increasing well-being of the Australian colonies, promising a great future, that the first settlement for many years depended for its very existence on supplies of food coming from England and that it was 'touch and go' should the supply ship be late. It is hard to believe also that all that was initially done with money voted by parliament. However time and human endeavour removed the obstacles and the discovery of rich mines is speeding now the growth of the colonies.

Many years will pass before Australia will take the rank to which the climate and the natural riches of the land entitle her, because the population in relation to the huge, still largely unexplored country is very small. Even today Australia is a queer mixture, on the one hand the dark-skinned natives camping in the forests, giving a stamp of wildness to the country, and on the other the drive to get rich, to penetrate the mystery of the land, to taste adventure which brings in a variety of people from nearly all parts of the world, people of different temperaments, talents and aims.

Seventy years have passed since the first British colony was founded on Australian shores and for the same number of years attempts have been made to civilize the native population with drink, tobacco and an assortment of diseases, all imported from England. The teachers had very diligent pupils. Natives in contact with the mines drink like regular John Bulls or Tipperary boys. Both men and women smoke with great gusto: a stick of tobacco is sufficient to gain friendship amongst men and favours from women. It is cheap, but then the ladies do not wear lace petticoats. The natives have even learned to cheat. I had the following experience on the diggings at Jim Crow in 1854.[1]

One day I remained longer than usual in my tent fixing damage after a storm. The miners leave early for work, but it was nearly eight o'clock by the time I finished the repairs. I brought a bucket of water and filled the dish for the dog, the guardian of my tent, and for a cockatoo, a blue one with a yellow breast. I swept the glowing coals of the fire together and covered them with ashes to keep the fire alive. Finally I put my odds and ends into a semblance of order. Then with my pick over my shoulder I started for my claim about two miles distant. The road wound above a deep gully through bushland. At one of the turns I spotted a blackfellow squatting with a *lubra* at his side. The natives throughout Australia are known as blackfellows, and they in turn talk about whitefellows —this covers all Europeans. Lubra in native language means wife or woman.

As I approached, both rose. He had a short shirt on of an indiscriminate colour, probably found abandoned at some old camping spot. I doubt if the shirt was worn for modesty but rather for vanity. It was tattered and just hung on his skinny shoulders with no attempt to button it, if such a thing were possible. Scars and burns, which replace the tattoos of other races, were clearly visible on his chest and shoulders. The lady was completely naked, and carried possum skins slung over her right shoulder, leaving her left arm free. She smiled, literally from ear to ear, showing large teeth, bobbed her head, shaking at the same time her curls which had never been combed in her life. (From time to time the natives rub fat into their hair and all over their bodies.) There they stood, one to the right and one to the left of me. The smell, far from the one usually associated with a lady using French perfume, reached me; I tried to superimpose the smell of my pipe on it.

Curious as to what was afoot, I moved ahead very slowly. Soon the lady pulled a clay pipe such as is commonly used in Australia from behind her ear. She blew into it, looked at it and showed it to me saying: "Empty." She was not disappointed in my good manners. Soon both of us were blowing clouds of smoke. Meanwhile the native began a conversation in broken English.

"Where are you going?"

"To work."

"Have you got money?"

"A little."

"Then lend me a shilling."

"I wonder when you will return it, if I lend it to you."

"What's the difference when?"

"Doesn't matter, here is a shilling. But it would have been better if you'd simply said 'Give me a shilling', because where would you get money? You don't like digging or doing any other work."

He looked at the shilling, said "Silver" and clenched it in his fist. There was no other place he could have put it. He spoke again:

"I'll stay here and you take my lubra for a walk."

Pleasant walk, I thought and said "I don't want to." He looked at me sceptically. There was an interval during which we all walked in silence. Apparently some new conception was forming in his mind. Finally he said:

"You know what, give me another shilling and I'll bring you gin."

"I don't drink."

"What sort of man are you? Drink is good."

"Just as good as a walk with your lubra", I said laughing, and turning off the road, went off to my nearby claim, never to see the couple nor my shilling again.

There are barely two million people (less than London) in all known parts of Australia, and this is not enough to change a land the size of the whole of Europe.[2] Also the British colonial system acts more like a brake than a help in colonizing. The system orders partitioning of the bush into sections, which are sold at high prices—one acre for £1. In places where future towns are to be established the price is £4 to £8.[3] Some individuals arrive in Australia with a large family and either have, or earn after arrival, a few hundred pounds. Should a section of land be granted to them on a ten-year repayment, such individuals could buy a horse and plough and clearing the section by sheer hard work, could save enough to feed themselves whilst building up the holding and paying for it. But if they have to pay £80 for a section, £60 for a horse, £20 for a cow and £40 for a cart, they have nothing left for the plough, the harrow and supplies until the first harvest. This doesn't take into consideration building sheds to live in and other small but necessary expenses.

The first colonists, the convicts, were given a few square miles of land free, but today an honest, hard-working man has to pay the government for a small plot on which it hasn't expended even the cost of a single charge of gunpowder for a carbine.[4] The first settlers were maintained by the government. For a long time food was imported in an attempt to enlarge the colony by any means. Now the government demands money for a bit of scrubland and no one cares how one manages to keep alive.

Admittedly, working in the newly-discovered gold mines one can earn a living and, if luck assists, more than that. So today there isn't the same need to care for the population as in the early days; but if the expansion of the Australian colonies is the government's aim, it should not charge high prices for pieces of virgin bush. There are very few newcomers to Australia who can afford the money to organize a farm, few who can earn enough money on the diggings for that purpose in a country where every item costs a lot of money. Land purchase becomes the prerogative of a wealthy minority,while the majority tramp through the bush from place to place seeking the gold which Lady Fortune—a capricious lady even in the antipodes—grants only the chosen few. The wealthy buy large tracts of land because there is no limitation on the size of their purchases, as there is in America, and let it on tenure to others. They make huge profits out of this renting, making in this new land another England; but I am convinced that it will lead eventually to Australia imitating America in its attitude to England. Already during my stay in Australia I witnessed riots and disturbances which I'll describe later. There are no factories here at present except for the manufacture of soap and candles. Australia has to import such items as food, clothing, implements and arms. But she has all the necessary potential for independent existence, awaiting only the development and implementing of this potential.

As soon as her inhabitants realize their strength, if they so wish they will get help from the Americans, who must at present pay a large duty for goods they import into Australia, which they resent. Business reasons and the dislike Americans feel for the motherland (England) will make them eager to help Australians. The numerous contacts I have had with Americans allowed me to fathom out their intentions as well as the attitude of Australians, particularly those of Irish origin.

The population of Australia has never been ascertained. Apart from settlers and people working in the mines, there are the wild natives tramping through the bush in groups of a few, or a few dozen, occasionally in bigger bands, without a fixed place of abode.[5] Without houses, huts or tents, they rest under the trees—here today and gone tomorrow. They leave their temporary camping places according to whim or the need to get close to rivers where it is easier to obtain nourishment, which consists mainly of kangaroos, possums, fish, wild duck, grubs and snakes.

They kill kangaroos with the help of a hardwood spear which they can throw accurately over long distances. They take a very curious stance to throw a spear. Standing on the right leg, they raise the left knee so that the hip is at right-angles to the body. They rest the spear on this hip and with the right hand move the spear a few times along this rest, estimating the distance to the target. Then the final throw, and the spear flies away from the hip to hit the target.

They pull possums out of the boles of trees. Their eyesight, better than an eagle's, notes the slightest sign of this small animal on the bark of the tree and they somehow know in which trees the animals usually hide. Natives have shown me those signs on several occasions, but the eye of a civilized man cannot see them. Their skill in climbing trees is exceptional. A hole small enough for one toe is sufficient to hold the weight of the whole body, whilst a similar hole is made higher up the tree with the aid of a small tomahawk. During this operation the native's free leg clasps the tree, and in such a fashion, as if glued to the trunk, he holds himself without the use of his hands.

Natives catch fish simply by inserting their hands in water, a trick their women can also do. But kangaroos and possums remain the man's domain. To catch large fish like the cod, which is to be found in large rivers, they use a very long hardwood spear. Lying on the high shore, or on a tree growing over the river, they wait patiently, sometimes for hours, until the fish swims by. Gently immersing the tip of the spear in the water to avoid frightening the fish, they gradually bring it near and give it the final thrust only when there is no doubt that they will hit the fish. I have been told that if they see wild ducks they dive under water and catch them by their feet, but I cannot guarantee the truth of this statement. They are so agile that I am inclined to believe that it is so.

They have remarkably well-developed hearing and eyesight. If there is need to solve a murder, to find the body and the murderer, the native is certain to do so. I can quote the following incident from New South Wales:

A settler left home one day and had not been sighted since then. It was assumed that he became lost in the bush, but the search for him failed. The suspicion grew that he was murdered. Eventually a group of natives newly arrived in the district was approached for help. One blackfellow took the search party to a small deep pond in the vicinity, pointed to a spot and said "Here". When asked how he knew the body was in the pond he replied: "The body has been in water for a few days now, otherwise I wouldn't know, but now I can see the body-fat on the surface."

"Where is the body-fat?" he was asked.

"Just where I have pointed out", replied the native. Although nothing was seen by the members of the search party, the net brought up the body.

Each family of blackfellows lights a small fire. They consider their white brothers fools because they make fires so big that they cannot get close to them. The black sits practically in the fire and bakes the possum, which he consumes, entrails and all. First the head of the family eats as much as he needs to satiate his appetite and then passes the meat to his lubra and children. Single fellows, known in the tribe as "poor boys", have their own fire and eat in an established order of seniority.

The natives don't have a common language. I have been told that tribes around Adelaide, in Victoria and in New South Wales have different sounding words for many objects. The same applies to their habits, or rather rites. Although they are so wild as to resemble in some respects wild animals, they do have a concept of ownership regarding their wives, but moral tenets are not very strictly adhered to.

As an example of their customs, in New South Wales near Murrumbidgee, a young woman is made to stand under a tree where the future husband approaches with a piece of hardwood which he places on her front teeth and hitting the other end of the wood, knocks them out. In Victoria this does not apply, because all the lubras I've seen have very large front teeth. I understand that before a wedding some secret rites take place on the lubra, connected with pulling out a lot of her hair.

Natives pay respect to talismans which consist of small, very clear crystals often found deep in the diggings but also to be found on the surface three miles from Bathurst near the Macquarie River. The method by which the *crodje* (medicine man) fools his client when giving the crystals is funny, although the preparations preceding this are fear-inspiring. On a dark, still night small fires are lit which burn very brightly because they consist mainly of bark. Near the fires dark figures painted with white and red clay jump and dance, screaming from time to time. Women sitting in groups hum a monotonous dirge. White and red painted black images jumping and moving in and out of the firelight and issuing terrifying screams form a macabre picture. Looking from afar the imagination tends to suggest some cannibalistic rites, particularly when two natives bring a stiff object, six feet long. But it is only a long piece of bark painted white and red with clay like the native bodies. The *crodje* gets up and walks slowly around the bark, examines it carefully from all sides, and sniffs it all over, apparently to sniff out the place where the talisman is. Although everyone knows that stones and crystals cannot be formed out of bark, many Europeans are willing to believe that the crystals are actually in the bark and not in the medicine man's hand. Eventually the *crodje* puts his mouth to the bark, sucks and spits into his hand and finds the crystal for the expectant native. He repeats the performance as many times as there are eager recipients for talismans, which are supposed to guard against evil.

So far no one has confirmed whether the natives believe in Almighty God. They claim that somewhere on top of some high mountain lives a bad spirit Yahoo, who devours people, particularly small children. This belief covers the crimes committed by the women who murder some of their children if those become a burden. When a child disappears they answer when questioned about the child, "Yahoo took it". It is impossible to trace Yahoo's steps because he can twist his legs to walk with the heel in front of his toes!

Natives fear thunder, and some claim that they believe in the possibility of the passing of the soul from one body into another. Wilkinson, who when living for many years in the bush as a station overseer taking care of sheep and cattle had regular contact with the natives, told me that they believe that the spirit of a blackfellow passes on death into a white man. I don't believe that tale because if it were true the natives would never kill a white man, and yet it often happens that a lonely shepherd gets murdered.

They are cunning but foolish, because they think they can fool a European by stupid tales. Let me describe an incident between Hugong the tribal chief and Rover, a member of his tribe who hated Hugong and wanted to usurp his place. Rover went to the overseer of a nearby station and asked him for a loan of two guns to shoot ducks, promising the overseer many ducks. He said that he would go with Hugong and asked the overseer to load one rifle with a bullet and the other with gun powder only. Rover went with Hugong and when they were alone told him he hated him and suggested that they act as their white brothers do and shoot at each other. Hugong shot first, but caused no damage, because there was no bullet in his gun. Rover shot next and Hugong fell down. Rover then returned to the tribe and told them that he had killed Hugong and was now the chief. All the tribe, including Rover, wailed loudly to show their sorrow over the death of Hugong. Rover was about to take possession of Hugong's wife when suddenly Hugong himself appeared and gave Rover a thrashing. The overseer had loaded both guns with powder only and the cunning Hugong dropped to the ground after the shot by Rover to make him think he was dead.

Some natives are not completely indifferent to clothing but they have no idea yet about what would improve their appearance. Possum skins were sufficient for them in the past but, trying to appear as decent citizens to their white brothers, they don a piece of an old shirt, or dress the lower part of their bodies leaving the upper part naked, and occasionally use a strip of an old pink woollen miners' shirt to wind around their heads. The women are loath to wear anything and even given a shirt wear it only at night for warmth and take it off in the daytime. The only exception I saw was at the station of Parker, a protestant missionary, where a single family of blackfellows lived in a wooden hut and dressed.[6] I doubt they lasted there long, because the natives do not take well to settled life.

There were attempts in Adelaide to tame some native women. Early in the history of the settlement, when there was a great shortage of women, a few settlers took native women as wives. At first the women liked the colourful clothes and the new mode of life but soon, forced to regularly carry out their household obligations, they began to pine for their previous way of life in the bush, or missed the regular hidings native husbands mete out to their lubras. In any case after a while the women disappeared, leaving most of the finery behind.

Some settlers managed to domesticate natives and use them for driving carts, looking after horses and similar occupations. They gave them decent clothing and plenty of meat and shelter under their own roofs. But not for long, for the natives, preferring bare bodies adorned by white and red clay, reverted to their original way of life. In any case they are not suited for hard work, and they cannot be used as protectors or guardians on long journeys, because they become frightened as soon as they cross out of the territory of their own tribe. The habits of the natives are such that when one of them crosses into strange territory, the whole tribe to whom the territory belongs seeks revenge on him and even on his tribe. The results are often large battles, luckily without serious consequences. The two tribes meet in the bush in a specified place, every warrior hiding behind a tree for protection, and for an hour or two they hurl spears and then part. It sometimes happens that a tree is not wide enough and some wounds result.

The natives admit that it is pleasant to have a roof over one's head and one of them was overheard saying to a pig in a sty: "You lucky fellow, you have your own *gunion* [a house, a cover, a shelter], you lie there all day and the white fellow even brings you your food." But in spite of their appreciation of the advantages, they are too lazy and prefer to get wet in the rain than build shelters. In all my four and a half years in Australia I only once saw a shelter, made from two long sticks and a few pieces of bark, with a few blackfellows under it.

They have a remarkable memory and never forget a face once seen. They have an unusual method of revenge. A tribe from near Goulburn River sent a message to a tribe near Bathurst that they wished to kill one of their members for hurting one of them. The elders of the Bathurst tribe decided that the guilty man was to be killed by his own son. The son speared the father but the father did not die immediately and until his death was tenderly cared for by the son. The funeral rites lasted a few days and then the Bathurst tribe went seeking revenge on the Goulburn tribe. In the battle a few Goulburn tribesmen were killed. Eventually the local police commissioner demanded that hostilities cease and a solemn meeting was arranged to make peace between the tribes. Armed to the teeth, the

Goulburn tribe went to the Bathurst tribe on a friendly visit. To stress the solemnity and importance of the occasion, English neighbours from a nearby settlement were also invited. The meeting was friendly and tribal dances were performed. Later, when members of the Goulburn tribe were asked why they came armed, the answer was that when they are armed similar visits end in a friendly fashion, but otherwise they sometimes end in arguments.

The armament of the natives consists of a spear similar to the hunting spear but shorter and thicker, called *wammara*.[7] Near the end are hooks made by notching the wood, or by using gut to fasten fish bones or hard small crystals to the wood. Such a spear is not thrown as in hunting, it is not supported on the left thigh. Instead it is inserted into a stick about two feet long with which the impetus is given. The forward part of the wammara rests on the left hand. So thrown, the spear is effective at about eighty feet. Another type of armament called *nulla nulla* is a club somewhat similar in appearance to an axe. The most interesting of all is the boomerang. It is made from hard wood, fashioned in the shape of a half moon. The method of using it is surprising. I don't know which end and how the native holds this thing, but after swinging it a few times in his hand, he throws it towards the ground a few steps in front. On touching the ground, as if pushed by a spring, the contraption rises high, and making a growling noise moves in a large circle to drop to the ground at the feet of the native. I have uselessly tried to imitate this throw many times, but have not been able to achieve any result. The last item of armament is an oval wooden shield. The natives are very skilful in repelling blows and spears with the shield, and this is probably responsible for the low percentage of tragic endings in their battles. They call all the arms by a combined name of *waddie*, but waddie also means wood for making a fire.

It is possible that other tribes in the unknown part of this large land have other arms and habits. Some of the miners who arrived in Australia several years before us and worked previously as shepherds or herdsmen far away from the mining districts tell of a taller and wilder race of natives who had bird feathers in their belts and adorned their spears in the same way. I have never seen those although I have covered a few hundred miles of the bushland. The inland, completely unknown, is covered by forests. The mystery of the land has as its two lines of defence huge distances and lack of water. Geologists claim that somewhere in the centre of Australia there must be a large lake, because some of the streams encountered do not flow towards the sea, but inland. All this is only a nebulous assumption, because a stream may change its direction a number of times and finally join with a river, or flow into the sea at some unknown place. (But there is some information that a lake does exist. Dr Leichhardt is mentioned as having spoken about a lake.[8] All this is word of mouth for I have never seen anything written about it.)

Just as I was leaving Australia, a fourth expedition was to be sent out.[9] The exploration was estimated to last two years. Only the future can tell if two years is enough time to cross Australia from south to north, and this only if the expedition does not meet the fate of previous expeditions. But one can expect better results from this expedition for it is to comprise forty men specially selected as suitable to stand up to hardship both physically and morally. The two brothers heading the expedition are energetic and far-seeing. They have brought camels, animals requiring less water than horses, to carry supplies. It is proposed to drive herds of cattle for food, but it seems to me that the half-wild Australian cattle will not be easy to handle on such an expedition. It is expected that they will move slowly forward from stream to stream, or to a well which they will dig with special equipment carried by the expedition. Australia is very poor in water. Many a time I could not find water on an eighteen to twenty mile march. Whatever the results, however, the expedition will expand knowledge about the land and water resources.

3

Melbourne and the road to the diggings

According to our agreement I am to be the cook. I rise at dawn and start a fire to boil water for coffee. Meanwhile the baker arrives with bread and a girl comes with milk well watered down, which she sells at higher prices to newcomers than to old customers. We sit on the grass for breakfast. All around are other tents under the trees where people breakfast and move about—future miners, merchants, drivers, farmers and even possibly a future member of the Australian parliament. Everyone tries to dress in the fashion of the miners. Overcoats, frockcoats and surtouts disappear and everyone becomes similar in outward appearance. Everywhere shirts in blue stripes, low crowned hats with wide brims, moleskin trousers, no neckbands or kerchiefs or ties. Legs used to patent leather shoes are now clad in half-boots or boots shod with iron plates at both ends and large nail-like studs in the middle. Everyone, animated by hope, discusses future plans and the sound of it all resembles a humming beehive.

I begin preparations for lunch, which is to consist of broth, meat and rice pudding. Stan, who speaks English best, goes to buy meat and rice.

"Why didn't you bring potatoes, onions, cabbage and some other vegetables? What will I use to make the broth?"

"I think you'll have to serve bread with the broth. All those vegetables are terribly expensive. I paid two shillings for four pounds of meat, but wouldn't have got a pound of potatoes for this money. Rice is cheap at three pence a pound."

"Well I'll have to stir some dough into the broth. Will you bring me an egg or two please."

But we had to give up the idea of using eggs to make the dough. Eggs were a shilling each.[1]

A few days passed while we made preparations for the trip to the mines, and meanwhile I visited the surroundings and the town. The rows of tents start at the foot of a hill fairly close to a bridge across the Yarra Yarra River. In front of us a marshy ground, behind the marsh a wooded hill. I mention this because four and a half years later I was fruitlessly looking for the marsh and the wooded hill behind it. Where the marsh was I found a wheat field; the green trees on the hill were replaced by white fashionable buildings of a new district of Melbourne called Emerald Hill. This is the speed with which Melbourne grows. One day surely it will be the size of London, if not larger. Even now it is a large town with wide streets and solid, good-looking structures, none of which is more than two storeys high. Everything breathes profit or the desire for profit. There are many established shops, and new shops are made from converted stables in which newly arrived merchants display their goods. I have seen a fellow traveller, who arrived with us on the boat, lease a sty, nailing boards to the walls and displaying ropes, buttons, pipes and other wares in the yet-unfinished shop.

"Why don't you wait until the shop is finished before you open it?" I asked.

"Every day costs money, so if I manage to sell some goods while I am lining the walls I won't have to use any more capital for living." A logical system.

The streets are full of carts and wagons pulled by horses and oxen. The empty ones return from the mines, the loaded ones are leaving town accompanied by groups of a few or more miners. A few lucky miners who struck it rich are in town to spend their money as fast as possible. Some of them walk about with a superior smile watching the new chums. Some, under the weather, drive in carriages accompanied by some well-dressed nymph. The carriage stops outside a jeweller's shop and soon the woman leaves the shop with a gold chain around her neck, a watch at her waist and a ring or two on her hand. But no matter —in course of time she will become a respectable matron.

Melbourne had started to develop as this view, done in 1856, shows. This engraving was made from a drawing by George Strafford and was originally published by William Inglis & Co. in 1857.

When one looks at the prodigality of these people, one brims with hope for one's own future. Here everywhere one can see gold, the basis for the licence and luxury of the Melbourne populace. The wrought gold as well as the nuggets displayed in shop windows are all the products of local mines. Perchance a lump meant for me waits in the ground and will help me to return one day to my homeland.

When walking along the streets sightseeing, I enquired about a laundry. My linen was badly in need of proper laundering after the long sea voyage. But the price of laundering a shirt is as high as the price of a new shirt in England so the only solution was to take my linen and a cake of soap to the edge of the river, where I found many others like myself.

Meanwhile my friends bought a good horse at auction to pull our cart to the mines. Official horse auctions are conducted twice weekly, where private owners entrust their animals to the auctioneer. For a fee of five shillings a head the auctioneer describes the horse, the age of the animal and the purpose for which it is suitable. If it is a draught horse, the auctioneer is bound to state what weight the horse could pull. If such an animal is not suitably broken in to pull a wagon, or if it cannot pull the weight guaranteed by the auctioneer, or even if it is discovered that it is a stolen horse, the auctioneer is bound by law to refund the money. But if the original owner recognizes a horse stolen from him after a lapse of a year since the horse has been legally auctioned, he has no recourse and the horse remains with the new owner.

Some horses are sold outside the auctions, but this is very risky, as more often than not the horse is stolen property. Some are sold in this fashion with forged certificates of sale from an auction place far away from Melbourne. The bulk of horse-trading is carried on by horse thieves, who are a plague, stealing from settlers and miners alike. They steal horses from paddocks, even from outside tents, and sell them in distant places to careless buyers or may even have an arrangement with some official auctioneers. Although an auctioneer caught selling stolen horses loses his licence and has to reimburse the original owner for the value of the animal, there are apparently some legal loopholes because this method of horse trading goes on.

Our good horse harnessed to the empty cart would not consider pulling it. We took it back to the auction yard and after considerable argument changed it for a mare. This one, suitable for riding, was not strong enough to pull our wagon filled with all the luggage and implements, and stopped on the first hill. We had a lot of trouble getting back to our old camping place near the bridge to decide what to do and when we got there we heard bad news! Two horses had been stolen from the tent lines, a woman had been drowned nearby by having a big stone fastened to her feet, and the ship *Cleopatra* on which were a few of our countrymen was now overdue. However the *Exact* also overdue, dropped anchor in the bay after 135 days' journey from Europe. Although it left long before us, it was so storm battered that it had to stop twice in various ports for repairs.

It was obvious to us that we had no hope of reaching the mines with our gear and only the small weak mare. So the following day we purchased another mare to help pull the wagon, and on 26 November left Melbourne.

The journey was not easy. The horses were used to English words of command, which we did not know, and so did not respond to our Polish. We travelled only about a mile and near the township of Flemington pulled off the road to prepare our camp for the night. Ours was not the first wagon there. Another one, loaded with bags of flour, with two horses tied to it, a drunken driver and his bulldog under the wagon, was already there.

Mine was the first watch. The moon was so bright that I sat on a chaff bag to write up my diary and to listen to the various sounds of the bush and to the sound of shots from far away—all so strange and so different. Well after midnight I woke my companion who was to take the second watch, gave him the two loaded guns and went to sleep.

Next morning we harnessed the horses to the wagon with the intention of starting off, but no matter what we did we couldn't make the horses pull. Our neighbour, sober now and in a good mood, lent us one of his horses to help pull our load back on the even road. Our mares then kept pulling—that is until we came to a hill where we unloaded half our belongings, but even so we could not make the horses pull. As we stood there in the middle of the township, debating what to do, a man approached us with the suggestion that we hire his horse to pull the load, and let ours walk behind until we reached the mines, where we could sell them at a good price. The proposition sounded reasonable so we asked him how much he wanted to take us to Forest Creek.[2] He looked around and said:

"My horse will pull the load but your cart is not good enough for the load you have on."

"How come? It's a new cart bought in England."

"Did you specify what load you wanted to carry when you ordered your cart?"

We did not realize that the size of the axles and the width of the wheels determines the weight one can carry. In the end it was decided to risk the journey. Our new-found rescuer told us that it was eighty miles to Forest Creek, that the journey would take the rest of that day and three more days, and that if the cart stood up to it he would get us there for £10 and the cost of the fodder for the horse. Eventually we struck a bargain to have the job done for £8.

He helped us load our cart, mockingly saying as he piled our six mattresses on the cart:

"Miners—mattresses—no future in this." He himself travelled without any mattress or blanket.

As we tramped at the side of the road, we passed parts where the bush was being cleared. We saw dwellings and fields of oats with large tree trunks still lying in the middle. The fat, well-kept cattle are proof that there is no scarcity of fodder. After three hours' march through this mixture of bush and cultivated fields we stopped at a dry creek, where only a few deep holes held water covered by a yellowish brown scum. And so to sleep because I didn't have to keep watch that night. I was woken by bird noises like the noise of our black grouse. But the noise-makers were birds called here "magpies"—like crows with black and white feathers. They always fly in flocks and are very noisy.

But it's time to get up and our Irish driver is impatient with the clumsiness of the new chums in loading the wagon. Within half an hour we pass a fast-flowing creek. What a pity we didn't know about it last night. We could not stop now after only a half-hour's walk, except to quench our thirst, but if we had pulled up here last night we could have had a swim and a good wash.

On the other side of the road was a small building with a sign "General Provision House". Two houses down was a hotel. These three buildings were a township called Keilor. We kept going in spite of the heat until late in the afternoon when we saw a lone house with a sign "Hotel". There was no water to be had there, only alcoholic drinks. We enquired of a few bullock drivers resting nearly and were shown a sheep station behind a clump of trees. We approached the well, but the mistress of the house demanded money before we could have water. We paid a shilling for a two-gallon bucket. But the water was so bad that even the horses did not want to drink it. It was no use stopping at this place so we continued on and after three hours found ourselves at the edge of the "Black Forest". Near the road stood a solid building with a sign "Gap Inn", but everything in inns and hotels costs money so we kept going. Eventually we pulled up on a meadow under the trees not far from some tents, hobbled the horses, and fetched water from some holes, which if not crystal clear was much more drinkable than the water for which we paid. From the people in the tents we bought some lamb and so spent the night there.

In the morning our impatient driver again had some remarks about our method of travelling.

"You are washing yourselves and packing like Prince Albert. Do you think you'll make a living in this fashion on the diggings? The sun is high up and we haven't packed yet."

Truly, what with the stove, the mattresses, the case with crockery, bundles large and small and a number of

S.T. Gill's impression of the road in the Black Forest. This lithograph was originally published in *Part the First. Containing 24 Sketches of the Victorian*

Encampment in the Black Forest. This engraving by Thomas Ham was originally published in *The Gold Digger's Portfolio*, 1854.

implements and tools, the bag of flour and other things unloaded, it was an unwieldy and long job to get it all stowed in again.

As we proceeded in the heat, the flies—a real plague in Australia—got in the eyes, nose and ears. It is impossible to keep them away. Some of our party have veils on their hats and have let them down as defence. I use a twig because I am afraid of infection. Australian flies are similar in shape to our own, although somewhat smaller. One species has a very virulent sting. If they barely touch the eyelid the eye stings and within twenty-four hours the whole side of the face swells, closing the eye. It has no ill effects and in a couple of days the eye is as good as ever, but the experience is not pleasant.

We keep going through the bush. Now on our right is Mount Macedon. It is a lonely mountain standing above ravines and forests, the summit nearly always hidden in clouds of mist and fog which practically every day come down as rain on the surrounding countryside. I have passed Mount Macedon on several occasions and every time was drenched by rain. The same happened to most of my acquaintances. But this time the weather was beautiful. Occasionally in the bush there was a lonely tent, put up by someone who had made enough money on the diggings and then had switched to a much steadier method of earning money, selling provisions and sometimes sly grog to passing travellers.

Spirits are officially allowed to be sold by public houses only after payment of a licence fee of £100 per year. Such a fee for having a long wooden barrack would appear enormous, if it were not for the talent of the English to drink huge quantities of spirit. Public houses can count on a high income only as long as gold lasts in the locality, but while it lasts the takings from a public house can be as high as $200 per day. So even a few weeks' business covers the licence fee and the cost of building the premises. It is a risky business because if gold peters out quickly the miners leave the field and the public house remains empty. That is one reason why the sale of spirits is restricted to the licensed premises. Owners of shops and

Diggers breakfasting at the coffee tent (sly grog shop). This lithograph by S.T. Gill was originally published in *Sketches of the Victorian Gold Diggings and Diggers As They Are* (Part 2), 1852.

tents caught selling spirits get their stock confiscated and pay £50 fine or are jailed for seven months. Still, the income from this sly trafficking is so great that those severe fines frighten no one, but only serve to make them cautious. Often a sly grog trafficker pays a £50 fine one day and the very next day is in business again, knowing it won't take long to recoup the fine.

Around eleven in the morning we pass the township of Bush Inn, comprising fifteen houses, quite solidly built for a new settlement. Barrels piled in front of one of the larger buildings indicate that it is a brewery. Our Irishman invites me to share a bottle with him, but I am more anxious to relieve my parched lips with water from the nearby stream. In this fifteen-house settlement, I found a Catholic priest and a blacksmith. Of the two, the latter is much busier and probably much better off financially, because he charges ten shillings for one horseshoe and large sums for fixing a cart. Travellers pay him a huge daily tribute because there are stony hills at both ends of the township and the shockingly bad road causes a lot of damage to horseshoes and carts.

Mount Macedon treated us unkindly, because a few miles out of Bush Inn we broke an axle. We had to unpack our gear at the side of the road, remove the axle and carry it back to the smithy. It only took one hour to weld the broken axle, but it set us back £2.10 and we had to remain for the night in the dark forest which dripped with rain and had a bad reputation for frequent attacks by robbers.

It rained all that night and till ten o'clock next morning, so we had a late start. The road was so bad that the horse and cart constantly sank in the mud. As there were to be four more miles of this bad road, we had to unload half our goods and proceed with the other half at least three miles to a locality called Five Mile Creek. We camped at the side of the road and a couple of our party took the cart back for the rest of our belongings. We only had some barley and some rice for our evening meal. Other travellers stopped from time to time at our fire and brought news that a woman was murdered in the bush not very far from our camping ground.

It was getting dark and we were anxious about our companions, when we heard their voices. Soon we saw them and the horses. The wagon was left not far behind, because the other axle broke. There was no choice but to stay. As we were comparatively rested and our newly-arrived friends dog tired, we had to go and find the cart and equipment in the dark and guard it through the night. Until today I don't know how we found it in the dark and the heavy rain. Fortunately I nearly tripped over a trunk of burnt-out tree and got the idea that it might contain some dry rotten wood. Soon, with the help of my tinder box, we had a small fire burning and enough light to find some more dead branches and dried leaves. We managed to make a big fire, which burnt in spite of the rain, but there was no place to lie down or even to sit down and as the fire dried our steaming clothes on one side of our bodies, the other side was getting soaked. Pessimistic thoughts accompany such situations—here we are with a useless cart, with horses no good for any work and not even sellable, and the smith about eight miles away. Who can carry a heavy axle eight miles through the bush? An awful prospect . . .

The dawn brought a spark of hope. The sun was out, promising a fine day. The early travellers advised us that we were only half a mile from Five Mile Creek and that the township boasted a good blacksmith. Our companions also arrived with some bread and tea and carried away the axle.

This blacksmith was apparently better than the one in Bush Inn, because he charged £4 for the repair. In a few hours our wagon was again fit for the road, but all our gear was wet and our mattresses waterlogged. Each one of them weighed more than a solid fully grown man and it would take at least a week of fine weather to dry them out. We had to decide to do without them. As it was it took us well into the afternoon to assemble everything in one place and load it up. We decided to camp where we were and have a well-deserved rest. We remained there the following day and one more night on the advice of

The main street of the township of Kyneton, looking north. This lithograph by S.T. Gill was originally published in *The Diggers and Diggings of Victoria As They Are* in 1855.

"Diggers licensing at Forrest Creek." This lithograph by S.T. Gill was originally published in *Sketches of the Victorian Gold Diggings and Diggers As They Are*, 1852.

our Irish driver, to allow the roads to dry and to fully dry our own gear to lighten the load.

Our two companions whom we had left to guard part of our gear when we went back to our broken-down cart, told us that during the night they were visited by a bushranger. The Black Forest is their main domain, because all the roads to and from the Bendigo and Forest Creek mines lead through it and thus all the gold sent from the mines has to pass through this district. Their story was that a well-dressed man galloped up to them and asked them what they were doing. He had pistols and a revolver in his belt. The conversation was difficult because our companions had very little English. The stranger asked about their nationality and when they said they were Polish, he trotted off. A countryman of ours had a somewhat similar experience tramping alone through the Black Forest. He met a stranger who proposed that they walk together. After a few miles and prolonged conversation, the stranger asked:

"Where do you come from?"

"Poland."

'Why didn't you tell me that before?" was the dismayed rejoinder.

"Why should you be so interested in it?"

"Because if I knew, I wouldn't have wasted so much time on you. It's well known that you Poles never have anything."

Not every traveller is so fortunate, particularly if he comes across a number of bushrangers. Sometimes they tie their victim to a tree and leave him to ants, mosquitoes and hunger. Very rarely is the unfortunate found in time, more often one finds a skeleton tied to a tree. Hardly anybody travels with gold. The few fortunate ones lodge their dust and nuggets in the camp with the local commissioner, to be carried under escort. A camp usually comprises a special commissioner to receive gold, a judge, and an officer in charge of both mounted and foot police. Special two-wheeled carts with four horses in harness, and assisted by a mounted escort, carry the gold to Melbourne. They always travel at the gallop, changing horses every two hours at special wayside stations. Even so, they were occasionally attacked and the gold robbed. The government is not responsible for gold lost in robberies or stolen from police camps.

Just as we were repacking our dried-out gear to be ready to resume our journey on the morrow, a patrol of mounted police arrived. Our compatriot, Leopold K., a police cadet, was a member of the patrol and stopped to talk to us. Tomorrow we would pass through the township of Karlsruhe, where a police training camp and an escort depot were situated. Our countryman was attached to the depot and invited us for breakfast there.

Next morning we travelled through a forest not so densely wooded as the centre of the Black Forest and in two hours arrived in Karlsruhe. After breakfast another hour's journey brought us to Kyneton, deserving the name of town rather than of township. It has only one street, but a large number of houses, a few butchers and bakers, a hotel and a shop selling the latest creations of European fashions.

Past Kyneton the environment is more open and the roads better. By midday we pulled up for rest in Columbine. I don't know if I should call the place a township —just a few houses, a muddy river, and a wooden bridge over it. We covered twenty-five miles that day and stopped for the night near a household where our Irish driver was well known. We decided to rest here overnight and the following day, and send one of us ahead to Forest Creek only five miles distant to reconnoitre and pick a suitable place for a permanent camp. Our companion went on horseback and soon returned on foot, having sold the mare, which was very fortunate as we needed cash to pay for the first month's licence of £1.10 each for permission to dig for gold. We blessed our good fortune, because without money to pay for the licence we wouldn't even be allowed to remain on the goldfields. There are regular inspections of the tents and those without licences are simply jailed.

4

Digging for gold

On 3 December then, seven days after leaving Melbourne, we triumphantly entered Forest Creek in an optimistic and enthusiastic frame of mind. But somehow our problems did not disappear, and as a matter of fact followed us wherever we went during our prolonged stay in Australia.

The road follows a gully. On both sides tents, dotted here and there, up and down, under the trees and near the sheer cliffside, right and left, without symmetry or order, varying in shape. They form whitish blobs set in the surrounding panorama, as dictated by the needs, whims or fancy of their owners. In front of some tents fires, and near the fires men, all dressed alike. Round hats with wide brims, blue striped shirts, moleskin trousers, shoes or boots heavily soled with metal studs. Long beards indicate that either there is no time to take care of one's appearance, or that no one cares about it. They stand in groups talking, or sit or recline on the turf or sand. It is difficult to read the expression in the overgrown, sunburnt visages; all seem to be indifferent, gloomy, absorbed. At first sight one might think they are bushrangers rather than men working hard underground. But one must not be influenced by outward appearances, we shall soon probably look just like the others. As from tomorrow when we begin work, we shall be entitled to call ourselves miners.

A stream meanders at the bottom of the gully. Nearby miners dig, mix something in vats, rock something in cradles. Over there a man squats right over the stream and shakes a large metal pan. What is he shaking? What are the others mixing? It must be gold, but why shake and mix and rock? We have read in Europe that it is pulled out of holes in the ground in lumps.

Holes are to be seen everywhere, round ones and square

Entrance to Forest Creek near the "turn off" to Bendigo. This engraving by Thomas Ham was made from a drawing by S.T. Gill and was originally published in *The Gold Digger's Portfolio*, 1854.

ones, wide and narrow. Near some of them poles—like one sees by water wells in Eastern Europe, near others winches with handles, built right over the holes. Each contraption has a hook at one end of the rope and pulls out baskets of soil from the holes. Some miners heap this soil to the side, others fill long bags with it, others carry it directly to the vats. What does it all mean? We'll have to learn it all, but it seems to be easy. It couldn't be difficult or complicated to dig holes in the ground, bring the dirt to the surface and rock it in a cradle. After all, every one of those miners we see around was a new chum once himself.

It is fitting to mention here that the first inkling about the existence of gold in Australia was reported to the British by our compatriot Strzelecki.[1] Several years ago this savant, engaged in a geological survey of New South Wales and Victoria, came to the conclusion, correct as the future has proven, that both colonies, but particularly Victoria, contained large quantities of gold.

This was a godsend for England, suffering from a surfeit of manufactured goods, seeking export markets to gain currency with which to buy grain for the rapidly expanding population. Already there are shipments twice every month to Istanbul and Smyrna, containing razors and knives. Turks, Bulgars and Armenians take those goods, even if they don't need so many knives and razors, and pay for them in the fruit of their rich soil, raw wool, raw silk and similar products, which are converted in British factories in turn into finished goods for export.

India and other colonies absorb the bulk of these goods. China is pressured into buying British exports. America refuses to accept the imports, since they produce many articles superior to the British. New Scotland, Canada and Newfoundland accept imports, but can pay for them with fish only. Europe doesn't absorb many British goods. The situation is drastic in spite of the Corn Bill and Reform Bill until a prophet by the name of Paul Strzelecki indicates that goods can be sent in large quantities to the new colonies of Australia and be paid for in pure gold. And gold can buy sufficient bread to support the Crimean armies.

Strzelecki crossed the Blue Mountains, calling one of the highest peaks "Kosciuszko" in memory of a Polish hero. This mountain is located near the border of Victoria and New South Wales, some fifty miles away from the sea. Until then no one had any idea that gold could be found in Australia. Some twelve years ago the copper mine at Burra Burra was opened.[2] The copper there is of high purity, about 90 per cent pure, but lack of hands and consequently high wages prevented the mine from being competitive.

Only England and Germany were sending workers out of their countries as the population was increasing out of proportion to the local need for extra hands. Hanover paid the total fare to Australia for anybody willing to migrate, demanding only a letter with a promise that the amount would be repaid when the recipient earned enough money in Australia. However there weren't very many volunteers for migration to such distant lands.

In 1839 and 1840 Strzelecki was conducting his geological explorations, and his report about the certainty of gold deposits in Australia was sent to the British Government on 2 September 1840 through Governor Gipps.[3] Some people claim that at the same time, or possibly earlier, Edward Hargraves announced his discovery of gold in Australia.[4] Certainly the Poles were never given the credit. However it is certain that until 1840 no one had an inkling about the existence of gold in Australia.

After that, investigation went at a slow rate until in 1849 and 1850 the European newspapers elected Australia as a fit competitor to California. The first find was in New South Wales and then, as some claim quite accidentally, someone found a rich place in Victoria near the spot where the mines of Ballarat and Forest Creek are today. Bendigo was supposed to have been discovered by an ex-convict—a once-famous English boxer, Ovens.[5] Thirty miles from the River Murray was the next discovery, practically on the Victoria-New South Wales border. Those were the main mines, with many side branches running in all directions from those points. After my arrival in Australia many new places were discovered, such as Mount Couron,

Jim Crow, Maryborough, Alma, Fiery Creek and many others.

In some of those mines work is very difficult because gold is to be found under a layer of iron oxides and the earth is so loose that the shafts and stops have to be lined with wood to avoid subsiding. In those places shafts are from 50 to 150 feet deep, sometimes deeper. Often they overflow with water which seeps through porous soil, as for example in Ballarat, where steam engines have to be used to pump water out. To operate mines like that the miners form companies of ten to twenty to keep up shift work in pumping out water, which not everyone can afford to do with steam-driven machinery.

The lining of the shafts is not very efficient in most cases, and small wonder, since it isn't done by experts. The miner is more likely to be an ex-soldier like myself or possibly even a priest. In spite of knowing what to do, the practical training is important. To cut the boards, to fit them and tie them to make the shaft waterproof and to prevent the soil from subsiding, is a difficult art and a very expensive business when over a hundred feet of shaft has to be lined.

Sometimes the company becomes so broke that the members haven't enough money to buy bread, with the shaft still not dug to the required depth. In that case either they take in new partners with money to carry on the digging, or if the shaft is in line with other gold-bearing shafts and there are good prospects of getting gold, they sell a number of shares in their shaft to raise the necessary money. It is true that the more shareholders to a shaft the smaller the return per person and sometimes even a loss from the venture when all the expenses are taken into consideration. However with a large company for a deep shaft, where not everyone can be employed all the time, the more far-seeing ones spend their spare time in digging shallow holes and washing them in the hope that possibly some traces of gold will be there.

There is usually more gold in the deep shafts than in the shallow ones, particularly around Ballarat and in a place called Eureka. But this cannot be considered a fixed law, because large quantities of gold and even nuggets have been found in both deep and shallow diggings. Likewise both deep and shallow diggings often give no return.

Miners prefer deep shafts on the whole, because when they are gold-bearing they last longer and give a better, longer chance of a steady return. There is always the hope of a fortune at the bottom. Less fortunate miners pack their gear and tools and move to other localities, hoping to earn just enough to make ends meet. This they usually find in the shallow holes where one, or at the most two miners work, and where there is no need to line the shaft. If the soil is not too hard, one day is sufficient to get the shaft dug. A man working alone cannot get below ten feet, because it is a sheer impossibility to throw soil out of a shaft over ten feet deep. I usually stopped at eight feet when I worked alone. It is not only the exertion of throwing the earth out, but also danger from the soil and small stones rolling back into the shaft unless they are moved well away by a helper.

If it is necessary to dig deeper than twenty feet, a digger working on his own forms platforms part way down, where he can throw the soil from the lower part of a shaft which is more than ten feet deep. Once this platform is filled with the dirt thrown up, the digger has to come up to this level and clear the earth from there. This is a very back-breaking job, because the shaft has to be proportionally larger than would be normally necessary for a shallow shaft.

The shafts are usually dug three to three and a half feet in diameter for round shafts or two and a half by five feet for rectangular shafts. The round ones are easier to dig for there is no need to square the corners, but they are more difficult to ascend and descend, because footholds hollowed out in the sides of the shaft are a bit too far apart for comfort, particularly if rain should partly wash those toe holds out or make them slippery. In a shaft only two and a half feet wide there is the possibility of leaning against the other side of the shaft during the ascent and descent and consequently they are safer. If a shaft is too wide for safety a rope fastened to a nearby tree or to the winch acts as a safety measure.

In very deep shafts however, such as predominate in quartz veins and which sometimes go two hundred feet deep, leather bags are used for lifting the dirt and the miners move up and down sitting on a piece of board attached to the winch rope This saves a lot of time in getting down and coming up again. When the gold-bearing stratum is no more than two to seven feet deep and the soil very loose and powdery, the whole of the surface soil is moved away. This is called paddock mining.

According to the latest Australian regulations, a claim for an individual miner is twelve feet square. This is marked out with pegs in each corner of the claim, the soil is turned over between the pegs thus forming the square, and the shaft is dug in the middle of the claim. However if a miner finds a lode where no one has ever dug before he is entitled to peg a claim two or three times larger. Where the diggings are poor no one bothers much if the claims are a few feet over the regulations. But on a rich lode not feet, but even inches can be the occasion for arguments, fights and sometimes murder.

Every claim has to be worked at least a few hours every day. If no one works it for twenty-four hours it can be claimed by someone else. Some miners try to mark a claim while they work out the old claim, particularly if there is a new find in the vicinity. This is difficult for one man or a two-man team where only one hour a day can be spared on such a new claim. The new neighbours soon wake up to it and throw their tools out of the shaft and take over, because the mining law states that a company of two miners cannot have more than one claim. This causes many conflicts, because the commissioner cannot be present near every new find to adjudicate.

When I was working a claim at the Alma diggings my claim was jumped by a Tipperary boy. I assured him that my iron pick was very sharp and that I never look where I use it once I am in the shaft. A little encouragement from my neighbours convinced him to move out. Waiting for a commissioner to adjudicate a dispute is costly, because a claim cannot be worked until the case is heard by the commissioner and four assessors selected from among the

Commissioner's tent & officer's quarters, Forest Creek, Decr. 1851. This engraving by Thomas Ham was made from a drawing by David Tulloch and was first published in *Five Views of the Gold Fields of Mt. Alexander and Ballarat in the Colony of Victoria*, 1852.

"Marking the claim". This lithograph by S.T. Gill was originally published in *Sketches of the Victorian Gold Diggings and Diggers As They Are*, 1852.

local miners. Therefore quite a number of disputes are solved by fisticuffs or more gory means.

I have previously mentioned "bottoming" a shaft. Bottom is a term describing the earth level at which gold is found. There are three types of bottoms: a white soft clay called pipeclay; a yellow sandstone; a stone which splits into fine sheets called slate. If on reaching any of the three types there are no traces of gold, it is useless to dig any deeper. There will never be any gold there. But one has to know the types of bottoms, particularly the pipeclay, because there is another type of clay, which after exposure to air for a few hours resembles pipeclay. Pipeclay proper is greasy and doesn't break away easily. Many a beginner abandoned a good shaft when coming across this false pipeclay and didn't find any gold. The slate formed into perpendicular layers contains some gold between slate sheets so the shaft has to be dug through it, because under it one often finds rich washing stuff—earth containing gold. The sandstone is the safest and easiest to work. All the gold, if there is any, is on top and the sandstone forms a good solid bottom to the shaft.

Above this bottom one finds the washing stuff—gold-bearing earth in all colours: grey, yellow, orange, red mixed with olive, light green on MacIvor fields, white, but never black. It is mixed with sand and stones. Sometimes the gold-bearing layer is only an inch thick, occasionally it deepens to a foot but even then it is usually best near the bottom, because gold in alluvial formation sinks down because of its weight.

The miners often leave the upper layers of the washing stuff in heaps near their tents, waiting for winter rains, when a lot of earth can be easily washed. The richer, lower layers are washed immediately, even if it means travelling a few miles to a stream. Occasionally large nuggets can be found six to eight feet from the bottom. The gold and stones there often looked as if covered in soot. Apparently some volcanic action placed this gold so high. Therefore, on diggings near volcanic formations, it is wise to test for gold over all the different depths.

One of my compatriots went into a worked-out claim

at Maryborough and feeling the earth around, discovered a nugget weighing four pounds, eight ounces, two penny-weights, valued at £224. It was found seven feet above the bottom of the shaft. The lump was the size of a fist and covered with soot.

The miner's main tool is a bar sharpened at both ends. One end is sharpened to a point, the other like a hoe one inch wide. The pointed end is used to break up gravel and becomes blunt very quickly, sometimes two or three times a day. It has to be sharpened by the blacksmith and the charge is sixpence for sharpening one end. When a new digging is discovered the blacksmith and the butcher are usually the first to arrive. After the butcher the blacksmith is the most important man on the diggings. The other necessary tool is a spade, and when one has to break through oxide of iron deposits, steel wedges or chisels are necessary. This is slow, awkward work.

Two buckets serve to bring up the earth. A crane or a winch with a rope attached to a metal hook lifts the buckets out. The miners change about, working above and below the ground. The work below is very difficult and nearly always in a crouching, bent position because of lack of space. Usually it is half a day below, half a day above. Below it is very hot and the air is stale. One way to overcome this is to make out of cloth a chimney eight of ten inches in diameter. The lower end is near the miner's head; the other end some feet above the ground has a sort of side funnel which can be positioned into the breeze, to force fresh air below.

It takes from several days to a few weeks to arrive at the bottom. This depends on the depth of the shaft, the type of surface to break through and the presence of underground water which may flood the shaft unless it is properly lined. When the miner has some indication that he may be nearing the bottom, work slows down, the digging proceeds in one corner of the shaft only, to facilitate watching for slate or sandstone, because fine particles of gold or gold sand are practically invisible in the shaft. One has also to be careful not to mistake mica, which shines, for gold.

When one of the three kinds of bottoms is reached, the miner calls out to his partner above ground: "Washing stuff". So warned, the partner places the bucketfuls of this earth on a separate spot until it is all collected. The miner underground scrapes the bottom with his knife, removing from corners and slight depressions all traces of gold. This is the testing time, when the miners find out whether the washing stuff contains sufficient gold to pay for all this back-breaking work, for the cost of tools, for food and a little bit over.

The washing is done as follows: four or five bucketfuls of the washing stuff is put into a tub, the tub filled with water to the brim and the whole stirred until all the soil is dissolved. Then carefully and slowly all the dirty water is poured out. This operation is repeated time and time again until all the mud is poured out and only stones and gold remain on the bottom of the tub. A cradle is used for further washing in conjunction with a hopper which fits over and above it. The final separation of gold is done in a tin dish, where gold heavier than sand or stones settles on the bottom. If the results are not very encouraging but there is a lot of washing stuff, a "Long Tom", a magnified version of the tub, requiring a lot of water, is used, particularly in winter.

If the results are encouraging, the work proceeds on the whole site of the claim by digging sideways from the shaft. The tunnels are dug as shallow as possible to save labour and the miner works squatting or sitting by the light of a candle. If poisonous gas accumulates, which causes the warning candle to flicker and go out, a fire is made from leaves in the shaft. This blows the dangerous gas out and work can proceed. In any case this tunnelling is a very dangerous occupation, because sometimes large stones above the miner become dislodged and may fall on the miner. Tired and semi-hypnotized by his monotonous work, the miner carries on, deeper and deeper into the bowels of the earth, always in the hope that the means to become free of this work await him somewhere there in front of him.

It is wise to work towards the edges of the claim and

remove the washing stuff from there and only later remove it from the central part of the claim. By doing this one avoids arguments and possible quarrels with neighbouring miners. As the side shafts are three to four feet wide there is solid earth between them, which serves to support the workings. To eventually remove this earth as well, it is necessary to build wooden supports into the side shafts.

I described a method of washing for gold, but there is another one called "puddling". This requires a large supply of water, at least two horses, and four men. The principle is similar but because of its size it is suitable for washing poor stuff and even surface soil from the mines.

Two miners using a tub can wash up to 150 to 160 buckets of stuff daily. The "Long Tom" with three men can deal with 300 buckets daily. The puddling machine can handle 600 to 700 buckets per day. If the puddling machine is located in a place with a good water supply and plenty of surface stuff, one can expect four penny-weights of gold worth sixteen shillings per 50 buckets, which at 650 buckets per day gives a return of £10.18. Allowing for feed for two horses and some expenses, this leaves about £7.10 per day to be divided amongst four men. Occasionally one can do as much as 1,000 buckets per day, which could make allowances for a poorer yield per bucket.

These methods are used for alluvial gold in its various forms from fine sand up to large nuggets. The largest lump I was allowed to handle was in Maryborough, found seven feet down by two ex-convicts. It weighed fifty ounces and was valued at £4,150. The lucky miners sent it to Melbourne and soon followed their find in person, but they were back in ten weeks, having spent it all riotously. An even bigger nugget was found in Ballarat in 1852. It weighed 1,644 ounces. Maybe even larger ones await a lucky finder.[6]

The quartz mines are no less bountiful than the alluvial ones, but quartz with veins of gold running through it is found at much greater depths and nuggets are not found in this type of mining. The shaft digging is similar to alluvial mining, but the shafts are much deeper. When

lling". This lithograph by S.T. Gill was originally published in *Sketches of the Victorian Gold* *ıgs and Diggers As They Are*, 1852.

I returned to Europe from Australia, I left at the Maryborough Quartz Hill a shaft 208 feet deep, which had not yet reached the gold-bearing strata and which was being flooded by an underground stream carrying water resembling sea water in taste. The miners were negotiating to bring a steam pump from England, which was a very costly undertaking and also difficult, considering the distance from Melbourne over most atrocious roads.

Consequently the income from a quartz mine must be correspondingly high to compensate for the preliminary expenses. I bought a tenth share in this claim for £100 and had to invest a further £50 as my share for wages and other expenses. However in spite of a careful inspection of the shaft and the fact that I saw gold specks in the quartz, I lost my money on this unlucky venture because the gold vein did not go further down and the whole undertaking turned out to be unprofitable. Some other claims got up to thirty tons of gold-bearing quartz averaging twenty-five ounces of gold per ton. Some poor shafts return only three to four ounces per ton.

Quartz is a hard white stone with specks of gold through it. It is a beautiful spectacle, but hard work. After digging it out, it is first burnt in a wood fire and then crushed in machines into a powder like flour. There are various types of machines for this purpose and the finer the powder, the easier is the separation of gold from it. The separation is achieved by the addition of quicksilver to the wetted powder. The mixture is placed on a tin dish and the sand is washed out just like in alluvial mining. The residue is placed on a piece of deer skin and the quicksilver squeezed out of it. The remaining gold is placed in a retort where the rest of the quicksilver evaporates and the small particles of gold are melted into a bar.

The fact that some quicksilver is lost in the process and that the crushing is not always sufficiently fine and effective makes this method expensive and only partly effective. During my time an effort was made at Sandy Creek to erect an oven to melt the quartz and from press reports I understand that further attempts to improve on present methods are being made.

Gold is where one finds it and there are no hard and fast rules in selecting claims. It is usually easier to find it at the bottom of a valley near a stream. A shallow shaft is usually sufficient then, because gold, due to its weight, tends to find the lowest point. Deposits usually run due east-west, which also helps in the selection of a claim. Usually gold is most plentiful near quartz, then granite, then iron and so on.

5

Diggers and mates

It is more than fitting to touch upon the life and habits of the miners, mainly because this very large society comprises men from all parts of the world, all countries and religions, varying dispositions and education, all types of artisans, artists, literary men, priests, pastors and soldiers, sailors, wild tribesmen with tattoo markings and those deported for crimes—all mixed into one society, all dressed similarly, all forced to forget their previous habits, leanings, customs, manners and occupations. All forced to follow their new occupation and to live the monotonous lives of the miners.

As they dig shafts next to one another, their outward appearance does not signify their previous importance, worth or mental attainments. A colonel pulls up the earth for a sailor; a lawyer wields not a pen but a spade; a priest lends a match to a Negro's pipe; a doctor rests on the same heap of earth with a Chinaman; a man of letters carries a bag of earth; many a baron or count has a drink with a Hindu, and all of them hirsute, dusty and muddy, so that their own mothers would not be able to recognize them. Many a one would not, a short while before, bother to look at a fellow with whom he now works. Here we are all joined by a common designation: "DIGGER". Only various shades of skin colour and speech denote nationality and origin, but it is impossible to guess previous station in life or background.

From early morning work goes on here above and below ground. Lifts and cranes squeal and some technical English expressions, used by everyone on the diggings, meet the ear as the diggers call to their helpers above ground. One hears "bucket", "aye, aye", "pull up" and so on. Everyone is engaged in the same occupation, uses one language to utter these short staccato words, because whoever wishes to work has no time for long discourse. But when the hour of rest approaches and the workers crawl out of the shafts to quench their thirst, take a deep breath of fresh air and relax with a pipe, then chance, sympathy, or some unknown attraction brings together total strangers and causes men from neighbouring claims to mingle. People who never knew each other sit or stretch on the ground and familiarly and freely converse, discuss matters, or bring up problems requiring solution.

Then one may well look with surprise at the ragged individual who shines and supersedes all others in his manner of speech, his wit and wisdom. If, attracted by those gifts, one enters into a closer acquaintance with such a person, one is rewarded by erudition and culture where the appearance of the person would not indicate such gifts in Europe. But small wonder, because quite possibly this uncombed and sweaty forehead was once adorned by a doctor's biretta, a church mitre, or a coronet.

Conversation is not always on weighty and serious matters. Sometimes a happy-go-lucky German tailor, a brawny English smith, a slightly-built French cook, a Polish Jew, an American or Dutch sailor, watchmaker, confectioner, a Swiss hatmaker, an impoverished Spanish hidalgo, gather near a mound of earth and one can see amongst them here and there a black Negro head, a brown Hindu face or the olive countenance with slanting eyes of a "child of the sun". Elsewhere in a group a Swedish sailor away from his whaleship, a Norwegian reindeer herdsman, a gaucho from La Plata, a Creole from Malabar or Mozambique and many others sit together. They do not debate matters literary, fine arts or politics. They amuse themselves with conversation about their countries of origin and its habits, and describe events they have experienced, because every one crossed many lands and many a sea before arriving in Australia.

Listening to those stories, one can spend time pleasantly and learn many strange things because many of the story-tellers are witty and some even wax poetical in their descriptions of events. Sometimes they argue, and here the characters and dispositions of people so different in origin, shape and colour of body, upbringing and education come to the surface. Some watch all this kaleidoscope of happenings with just a passing interest as a temporary amusement which may leave faint impressions of pleasant or unpleasant memories. For others this is a source of important conclusions and psychological insight. Maybe some Knigge[1] is at present in Australia; maybe he is now making notes in his diary when the voice of his claim mate rising from the shaft shouting "Bucket" breaks into his scientific work, the diary goes inside the coat and saying "Aye, aye" he grasps the winch handle. I am sure that in time more will be told than I can give in this superficial description, but it would be wrong to try to say more. I write as I see it.

The miners' dwellings are tents of various shapes and sizes surrounded by a drainage ditch. They are haphazardly placed on the fields according to the miner's fancy, but stores, public houses, butchers and bakers usually put their canvas structures in one line and form a street, which is named "town" by the miners who camp around. On large goldfields such as Forest Creek, Bendigo, Ballarat or Owens, apart from this street of the traders and the miners' tents, one also finds a canvas chapel. Preachers, however, as missionaries, often have a habit of preaching to the pious congregation from a stump, a cart, or a barrel. There is also a school for the children, who multiply very fast in Australia.

Sometimes, when gold is plentiful on a field and the number of miners large, somebody improvises a racecourse and even a theatre for actors who do not disdain gold and travel from Melbourne to wherever it shines, claiming however that they have travelled only to entertain the "most honourable company" of miners, though they well know that the society on the diggings is too mixed indeed to be called "honourable". Even Lola Montez arrived in Ballarat in 1856.[2] She did make a mistake in assuming that she could expect the miners to kiss the hand of the lioness who would wield the whip as of old. I don't know whether her charm had lost its power, or if the hand

Store at the Diggings. This engraving by Thomas Ham was originally published in *The Gold Digger's Portfolio*, 1852.

wielding the whip had got heavier, but she left the fields far from victorious.

Bands of Czech musicians travel around the mines and often one can hear a song accompanied by a harp issuing from a tent, a familiar song so often played on a harp, about love . . .

The tents are occupied by one, but more often by two miners; sometimes by a greater number, occasionally by a whole family of husband, wife and children. In the tent are beds which are not purchased in furniture stores—every miner makes his own. At the end of the tent furthest from the entrance and at the longer side, the miner fixes four stakes on which he supports a sort of hammock made from two poles the length of a man and kept apart by cross pieces to the width of a man's body. On this he slips two bags which act as a mattress. At the head of the bed on a couple of stakes hammered into the ground, a solid piece of bark or a board from an old tub no longer suitable for washing gold, or a piece from a broken case in which wine, brandy or jam came to Australia, fastened with nails, makes a table. Similarly a case serves as a pantry where bread and rice are protected from mice. Coffee, tea and sugar rest in the same place, but meat is hung high on a tree, or at the top of the tent pole, so that dogs can't get at it. So hanging it goes black but lasts longer in the air than if hidden somewhere in the tent.

What other furniture is there? Most likely nothing else, because the above suffices to satisfy the barest necessities. One can either sit or lie on the bed, on a rainy day one can place a candle stuck into a bottle on the table and eat there. In fine weather one feasts outside on the grass near the fire, particularly on cold days, when mutton which coagulates quickly must be prepared during the meal. There are no spare clothes, hence no need for a wardrobe. If a piece of clothing or a couple of shirts get in the way, folded up they serve excellently as a pillow with a blanket for cover. Why possess more chattels? If a guest arrives, he sits on the bed next to the host, or on a bucket brought in from outside, turned upside down and presto a chair is ready.

There is very little trouble shifting from place to place. The bags are pulled off the bed, the tent off its poles, and in an hour the house and furniture are all pulled up and packed. Only the stakes which supported the bed and table, and the tent poles remain as a skeleton. If work keeps the miner for a long time in the same place, particularly through the winter and the rainy season, a fireplace of stones or sods may be built near the entrance to the tent. The opening of this fireplace is towards the tent and the chimney leads out through the tent roof; a cask without a bottom is placed on top for a better draft. Sometimes the miners make the fireplaces from wood, but these catch fire very easily and could destroy the tent. I never had a fireplace and often have been in a lot of trouble to start a fire with damp wood during rain.

As with the dwelling and the furniture, the kitchen utensils are not of great consequence. A kettle or a small cauldron for tea or coffee, a frying pan, and for those who bake their own bread a large tin with a cast iron overlid. The cupboard consists of a panikin which serves as a spoon for all liquids, a knife and a fork. But not everyone can boast such a fine cupboard. Only the most neat and fastidious who do not relish using their fingers to pick meat from the frying pan boast of such an assortment. Less choosy ones place a piece of meat on bread and cutting off small bits, eat both together, hardly worrying that the fat drips all over their hands. Possibly they even hold their meat in their hand on purpose, to wipe their fingers on their boots after eating and in this fashion preserve the softness of the leather.

The English do not partake of broth, and some others have no time to cook it unless it's Sunday, so a spoon is unnecessary. One can do without a fork, using one's fingers or picks made from twigs. The tin panikin, frying pan and knife substitute then for all dishes, plates, cups, glasses, goblets, silver fruit knives and all table service, because one can eat and drink with the aid of just the three implements which do not require stands nor buffets.

The kitchen and cupboard are carried by the miner whether he goes to the diggings or on a long journey. The knife is in a sheath in the belt, or in his pocket, the panikin hangs at his belt, while the tin pot, threaded through the pickaxe, rests on his shoulder with the frying pan. On a journey he adds a blanket rolled tightly like an army coat, draped over the shoulder and tied together with a piece of string at the other end. In this fashion all goes together with the traveller. When one travels through a forest there is no need for an inn, there is one under every tree. One stops at will, or rather one stops where there is water to be found, makes a fire, puts the pot on it, places a lump of meat on the frying pan and fries it. After the meal and having given the scraps to the inseparable companion—his dog—the traveller gathers wood to keep the fire going throughout the night. When all this is attended to he makes his bed, that is he clears stones, roots and bits of wood from the spot where he intends to rest. He would gladly gather leaves to make his rest more comfortable, but leaves do not fall off the trees in Australia and so he must do without a mattress. A wooden block under the head, or perhaps his head rests against the trunk of the tree. He wraps his blanket around him and so sleeps on this, nature's sofa without springs.

Only the dog cannot find a suitable place for himself. He would like to rest as close as possible to his master and walks in circles as if he wanted to roll himself into a bundle. He crosses from one side to another, but everywhere hard stones cover the ground. He scratches at them and moves them away but one seems to be grown into the ground just at his master's feet where he has decided to rest.

Wait, poor fellow, you need help. The miner gets up, removes the stone and covers the spot with the corner of his blanket, which he reserves for his companion. Now both of us are properly settled down. Stay put dog, I'll add some more wood to the fire, smoke a pipe and meditate for a while . . .

The time and the place are pleasant, the fire glows, the silence of the night surrounds me, the ancient forest trees say their evening prayers in a subdued whisper. My thoughts fly to the far distant family home, faces of dear ones loom in front of me, the imagination presents them to me in familiar surroundings, I talk to them, but the images become more and more blurred, more and more nebulous. The eyes close, the pipe drops out of my mouth and I am oblivious to the world.

Sometimes I sleep peacefully throughout the night and the early dawn or limbs aching from this less than comfortable bed wake me, but sometimes the dog leaps up suddenly, barks and disappears like an arrow into the darkness of the night. I lift my head, lean on my elbow and make sure that my club is nearby. I never carry any other arms in Australia because it is difficult to aim and easy to miss in the dark with a firearm. Neither applies to a cudgel or a club. Meanwhile I hear somebody walking, approaching under the escort of the dog who tries to get a bite at the shins of the stranger. Then a voice is heard.

"Call off the dog."

"Who is it?"

"Me, a traveller. I'm on my way to the diggings at B. I got lost and since sunset I've been searching for the road. Thank goodness I've spotted you, 'cause I'm dead beat. Would you have some tea? I'm all athirst."

"Over there near the stump. The quart's nearly half full. Warm it at the fire."

"God will repay you. There is no need to warm it."

"Here is the sugar."

"Thanks."

He takes his panikin off his belt, fills it with cold, somewhat stale tea, takes a bite of bread and drinks. Supper doesn't last long. He lights his pipe and settles down for the night next to me. Tomorrow we'll travel together because I'm also going to B.

I see I have moved quite a step from the description of tents and utensils. Involuntarily I recalled tramping through the Australian bush. My thoughts were there, and my pen followed my thoughts, but the remembrance of such an incident belongs, I think, to the scene I was depicting, so I have not sinned overmuch bringing for a moment myself and my dog into the action instead of a multitide of diggers travelling without a cart, afoot and alone to the far-distant diggings.

Some may query how a miner supplies himself with the necessary provisions on such journeys. Usually one follows a track where every day or two one passes a station or a few houses of a newly-forming settlement. In all those places one can obtain meat, bread or flour. A dough cake can be baked in hot ashes if no bread is available. Tea, sugar and a pinch of salt are always carried in a linen bag. Some carry a gun and shoot cockatoos and possums on moonlit nights, which they bake on wooden skewers. Possums can be shot only at night, because they spend their days in holes in the trees. Only natives know how to find them in daytime.

Naturally, no courtyards divide the miners' habitats. Anyone can erect a tent, six feet beyond the neighbouring tent. Nor is there need for more space. Near one tent wall is a heap of shovels and pickaxes, at the other a bucket placed upside down with the frying pan on top, nearby the cradle for washing gold. Close to the tent is a dog kennel made from bark and, in the kennel, the dog chained up. If the miner owns a horse and cart, then a separate tent is put up for the horse or he is tethered to the cart, which is placed at the entrance to the tent and close to the dog. The dog watches over the miner's safety and his property when the miner sleeps, tired after his exhausting work.

Many a time the barking of the dog chases thieves and bushrangers away, wakes up the miner and warns him about approaching danger. Everyone in Australia knows that he is not allowed to enter a tent without first calling out to the owner and that the owner can shoot at anyone approaching too closely or entering a tent at night without announcing himself beforehand. Without the dog many a crime could have been committed, because a miner sleeps soundly after heavy work. Yes, a good dog is a priceless treasure on the goldfields. A criminal approaching a tent and confronted with loud barking turns back, convinced that the prospective victim, already aroused, may give him a very warm reception.

The tent is not permanently secured because it can be ordered to be moved if it is in line with the shafts in which gold has been found. Anyone can demand the removal of a tent if he wishes to dig in the place where someone else's tent is pitched, because such is the law. Actually it's a rare occurrence that someone should try to move a companion, since one considers the possibility of reprisals.

All day long there is peace and quiet near the tents, the fires die out and everyone is at work. But when the sun begins to set the miners return homewards tired and hungry and therefore in the bad mood which is usually the aftermath of empty stomachs. In silence some kindle fires, others bring water and fill kettles for tea, others melt fat and prepare steaks. The silence is broken by the crackling of the fire and the crackling in the frying pans. Sometimes, here and there, a dog squeals when brought to order by the cook's boot or a stick for getting too close to the frypan. The tin panikins filled with tea cool off in the breeze. Groups of miners sit on the turf. One divides the bread, or after cutting off his share passes the rest to his neighbour. The cook places the steaming frying pan in the middle and takes his place in the circle. As the hunger is appeased the silence is broken, conversations start, interlaced by jokes and laughter, sometimes a sad or happy song resounds from one of the fires. One hears murmurs like in a beehive, a sign that everyone has finished their meal.

The sound of shooting begins, at first single shots and then as the number increases it sounds like flanking fire. This sound lasts well into the night. The intention is to fire away the old cartridge, and load again to make certain that the piece will not fail if needed. Actually it is a superfluous activity if arms are kept dry. Shooting takes place more to let robbers know that there is a gun in the tent, alternatively to gain assurance that there is no need to worry about an assault.

Those nightly salvos always made me feel uncomfortable, because many of the diggers have had little experience with firearms and were as proficient in handling them as I would be if told to change a baby. I'd

Forest Creek, Mt. Alexander. This engraving by Thomas Ham was made from a drawing by David Tulloch and was originally published in *Five Views of the Gold Fields of Mt. Alexander and Ballarat in the Colony of Victoria*, 1852.

be sure to harm the baby: these dilettantes do not break their pistols or guns, but send the bullets in all directions, even into other miners' tents. Sleeping miners have been killed by stray bullets. I nearly had it happen to me in Bendigo. By the sound of the passing bullet I knew it was close to my head for it did not moan like a high flying shot, rather it whispered by and suddenly the sound ceased as the bullet hit the ground. A similar shot landed near my compatriot Mr S. on the diggings at Back Creek when he was having his dinner. He picked up the spent ball without moving from his seat.

After coming alive out of a hail of thousands of bullets, it would be unpleasant to be shot by a dilettante English hunter. Let him fire at cockatoos and possums, but not at people in their tents. It is even dangerous to go hunting with them because they could mistake a man for a possum, though a man is about twenty times as big and looks nothing like it. Such an accident happened when I was on the Jim Crow diggings at Wombat Flat, when two companions went hunting and one shot the other. Afterwards he maintained tearfully that he was aiming at a possum and had not the slightest intention of hitting his best friend. Such a pleasant life in Australia. One has the choice of being shot by someone who intends to do so, or by someone who has no intention of doing so.

The noise and shooting lasts a long time, then becomes more widely spaced, finally ceasing as it began, with occasional shots. There remains the sound of drinkers returning from public houses and stores. They sing and cannot find their tents, which is not so easy even for a sober person in a locality where suddenly a few thousand tents spring up without streets, placed all over the terrain. The drunks scream and call sober people out of their tents, asking them where they are and for directions to their tents. It is difficult to answer when one sees a stranger —a drunken stranger. They receive many uncomplimentary answers, which in turn are reciprocated by "God damn" with further compliments following. Finally the last voices cease. The drunks have moved off, probably to sleep somewhere under a bush. A dog growls occasionally. The diggings are asleep.

The silence lasts only a few hours on diggings where water is scarce, where one or two wells supply insufficient water for a few thousand miners. Those who can afford it buy water brought in barrels from places some miles distant and pay high prices. In some places, such as Mount Malaga, the charge is 2/6 a bucket. Those who cannot afford such a price get up in the middle of the night and form a queue at the well to fill a bucket. Those behind carefully watch the queue, because quite a line waits its turn, eager to get back and not happy at all if someone tries to get water before his turn. This also causes arguments and sometimes fights and so the night's silence occasionally explodes into a rumpus.

With daybreak smoke lifts over the tent lines, the frying pans sizzle and the aroma of frying fat spreads over the diggings—breakfasts are being prepared. They are neither elaborate nor elegant; nor do they consist of many dishes. The same is served as for last night's dinner—steak. Only the beverage is different, for a quart of coffee without milk replaces a quart of tea. As soon as the kettle is emptied and only the grounds remain on the bottom the kettle is filled with water again and put on the fire to extract the rest of the coffee from the grounds. This is coffee number two, dirty in colour, so thin and without aroma that the most refined nose would be at a loss to recognize the liquid for what it is. The miners call it "cafio". This "fio" at the end apparently is supposed to describe the thin consistency of the beverage. But it has great merits because it is a good thirst quencher at work. A similar procedure is followed with tea leaves if tea is served for breakfast, the only difference is that this liquid retains the honorary name of the original tea, quite undeservedly.

The miners do not eat a full meal at midday, for although a couple of hours break would no doubt be enough to fry a steak, it would mean that the frying pan would have to be taken to the work place and one would have to go to the butcher for meat. A man so tired prefers to lie down, rest his body and gather strength for more work. So tea is boiled and a piece of bread is eaten, with cheese if one can afford it, usually just bread. Lunch consists mainly of rest—on a heap of earth.

Such, then, is the life of a miner. Morning and evening steak, or for a change mutton chops, at midday bread. Coffee or tea number one morning and night, number two as a drink in between. And so the same, week in, week out, months and years. Sometimes on Sundays when there is more time and the utensils are available, one can make a European-style broth for lunch, make some dough, throw in a few potatoes and a cabbage, and lo there is a feast! Only miners from countries where soups are an important part of the diet cook this, but not the English. They prefer plum pudding on Sundays.

All week the miners dig, wash gold and work like bullocks, Saturday afternoon they usually remain home, but even there they have a variety of tasks. Wood has to be collected and prepared for the forthcoming week. The sound of axes and the noise of falling tree trunks comes from all sides. The English often cut trees without actual need. They do it simply for fun, because here they can indulge their liking for the fuel which is as rare as saffron in their own land and which they use at home only for kindling coal fires. There is also the task of washing clothes. Those without a wife or a housekeeper do it themselves.

On Saturday the partners divide the gold won through the week. Each partnership entrusts the gold from the daily washing to one of the members until Saturday when they get together to weigh the gold intake, calculate how much each is entitled to, and give him his share. On Saturdays also, everyone goes to the store and buys provisions for the next week, because sometimes the stores are far away from the tents and there is no time nor inclination to go for a walk when one has already walked a few miles to and from work. And so, although there is no digging on Saturday afternoons, there is plenty of other work.

Sunday is the only holiday in Australia. They do not recognize any other holidays except for Christmas and Easter, for which they allow only one day each. This is wrong because considering the hard work which prevails on the diggings, even two Sundays per week would not be too much.

Queen Victoria's birthday is celebrated by the miners with improvised illumination. It is a pretty sight when on the eve of this festive day thousands of fires blaze through the darkness of the night, all lit at once on a signal given by shooting from various directions. Large heaps of dry gum branches with leaves left on are assembled in front of every tent and lit on the sound of the evening discharge of arms. But this is the end of festivities; next day is work as usual. Before the beginning of the war in the East there were gaps to be seen in the illuminations because the American and French tents remained in darkness. After the coalition and an American journal had published an article apparently demonstrating the willingness of Americans to help the mother country, the illuminations on the eve of the Queen's birthday went on without gaps.

The manifestation of this new attitude was not without exceptions and not always sincere. When I asked, "Why don't you join up and fight for the glory of your country?" a Frenchman replied: "Ha, if it was against England I'd join immediately." Nor was the article in the American publication flattering. It contained a sentence: "We cannot deny our reluctance towards old England, for we don't like her, but will not allow her to be hurt and were she drowning, we would rescue her." Those fires somehow demonstrate a certain sympathy between the two nations, but I doubt that it will last longer than the smoke from those branches with dry leaves. I am just making assumptions based on what I have seen and heard. Only the future may show if they'll come true. Meanwhile, having described the diggings and the life of the miners, I return to Forest Creek, where I began my apprenticeship in the mining trade.

6

From Forest Creek to the Ovens and back

We arrived early before midday on 3 December on the spot selected for us by our friends on the diggings. Our Irish driver unharnessed his Punch and left for home. Some of us busied ourselves erecting the tent, while others harnessed Gypsy to the cart to fetch water and firewood. When the cart returned the tent was organized, food was cooking, and all that remained was to settle down, inspect the surroundings and begin work on the following day.

While we were busy doing just that a young miner, passing by and judging by our hardly-sunburnt faces that we were new chums, asked if we were German. "No", I said, "but we speak German." He was a Hanoverian by the name of Degenhardt, one of four brothers working on the field, and he offered to show us the rudiments of the mining trade. Over a cup of coffee he advised us to shift our camp, because he considered it too close to the road for safety and privacy and suggested that we should move next to his camp where it would be easier to explain the know-how of the mining trade. Moreover it would be company, particularly as a countryman of his by the name of Wilhelm Gayer was camped nearby with his Polish wife. We only managed to inspect the proposed site on that day, and shifted over the following morning. As all of our neighbours were at work, we did not get an opportunity to talk to them until the evening.

The Degenhardts had arrived in Australia seven years previously and originally worked the copper mines at Burra Burra, where they all earned enough to buy eighty-acre lots of land near Adelaide, build houses on their selections for their wives and, leaving them there, to go off to the goldfields in search of a fortune, or at least enough money for improvements for their farms.

Our kind neighbours gave us much useful information to make our work on the diggings easier, and Gayer altered our cradle for washing gold, improving on the London-made article. One of the Degenhardts taught us how to bake bread and decided that one of us should take on the role of cook, provider and guardian over the tent for a week and then be replaced by one of the others in rotation. Unfortunately my turn came first and I had the difficult task of feeding and looking after men still used to home comforts and varied menus, men unused to the simple, repetitious fare of the miners. Consequently I washed dishes and prepared food constantly. A miner is expected to eat out of the pan in which he cooks his food; we had, at the beginning at least, a whole chest of cups and saucers, soup plates, meat plates and other utensils and could not imagine that all the paraphernalia of a gentleman's table might be unnecessary. Even today as I write and recall that first week, trying to marry a miner's life and appetite to an ex-officer's standards and expectations, I shiver at the memory.

On the third day of my officiating as cook, a slightly familiar face passed slowly by. In spite of a great change in his appearance, I recognized a fellow traveller from England, a Mr Plevens. "What's the matter with you?" I asked.

"I've been in many parts of the world", said Mr Plevens, "and I decided to try the life on the Australian goldfields. I was certain I would find some sort of an eating house where I would be able to eat. But there is nothing like that here. I am sick and have a fever, I have been given a corner of a tent by some people I know, but it's dirty and windy. I was cooking my own food, but the meat here is so bad that I can't eat it. I'll just die here."

We kept Mr Plevens in our tent and after a week he returned to good health and went back to Melbourne to re-embark for England. Life on the diggings without hotels and restaurants is suitable for hardy types but not for people like Mr Plevens.

When the first week of our stay on the diggings passed without my companions finding gold, I went out to try my luck. The first shaft I started was soon flooded with water. We worked all one day with buckets, but couldn't keep the water down. We decided to give up the shaft and start elsewhere. This time we chose a place somewhat nearer to our camp, for the previous shaft was six miles distant. We also took Gypsy to transport our tools. While we worked Gypsy cropped the grass, but somehow disappeared in the afternoon. My companion went to look for her and found her towards evening . . . with a foal the size of a big dog.

The following day we finished the shaft and took some washing stuff out. That was the first time I saw a speck of gold the size of a wheat grain. The whole of the next day we washed the soil we had dug out. Being far from expert, as I found much later, we left the best washing stuff in the shaft and probably lost more than half the remaining gold amongst the stones and sand. As a result we recovered only three-quarters of an ounce of gold, worth £3.6.0 from the shaft. But then one has to pay the piper when one cannot play oneself.

Our early days on the goldfields were very difficult. Unused to hard physical work, to standing knee deep in water, to accommodating to changes of temperature from hot to cold and vice versa, I felt the difficulties very much. Christmas Eve was particularly difficult for me. Used to fasting on that day I did not eat, but went to the shaft on an empty stomach. There is a vast difference between spending an idle day after a good supper and spending it with a pick in hand. I hoped to forget hunger by working hard, but I felt so weak and giddy through the day that I had to give up any idea of further work in the afternoon.

Christmas Eve was a very sad occasion for us all. There was no way to keep up our traditional celebrations so we spent Christmas Day on talks, discussions and making acquaintances. The English have a different way of spending the day. They played quoits, talked, had jumping competitions and some drank.

After the holiday we settled down to hard, difficult and monotonous work. Gayer, our neighbour, suggested I quit Forest Creek and go with him to Ovens, about 210 miles

"Nuggetting". This lithograph by S.T. Gill was first published in *Sketches of the Victorian Gold Diggings and Diggers As They Are*, 1852.

away. The proposition was tempting because I would see more of Australia, meet other people, learn a lot about mining from Gayer, an experienced miner, and perhaps earn more there than the very poor return our company was getting. However there were some serious difficulties. I was completely broke and could not pay my way over to Ovens, nor keep myself there until we struck a lode. I was also bound to my present companions by a promise that we were all partners for the first six months after arrival in Australia. I told Gayer all that, but the good-hearted fellow said:

"Ask your companions if they agree. So long as all of you stick together, none will learn the mining trade, nor make a living. It's wiser for all of you to split. And as far as money for the trip, you don't need any. I have sufficient supplies and I'll stake you. When we find gold you'll repay me."

"How can you be so certain that we'll find gold? I may have to owe you the money for ever", I said.

"I am certain that we'll find it", replied Gayer. "Probably not as much as we'd want to find, but enough to pay your debts and some over. I know what mining is. You can work for a week or two and get nothing, then you find it. And the Melbourne papers report that Ovens is very rich. If only half of what they say is true, we'll be all right. But we cannot shilly shally. On the diggings the early arrival has the best chance. We must leave here in three days' time. Meanwhile I'll buy a horse and cart and you have a talk with your companions."

My companions readily released me and I advised Gayer accordingly. He was very pleased and planned a pleasant journey during which we would shoot cockatoos for broth and so have a change from the monotonous goldfields fare. He showed me a solid cart and a sixteen-hand horse called Jack, capable of pulling great loads. I also met the other member of our team, Christian Schpötter, an old friend of Gayer from Hanover, a very good worker (according to Gayer) who spent all his money on drink and gambling.

When I tackled him about the life he led, Christian told me a sad story of orphaned childhood, always hungry and cold, working for a very mean peasant couple as a goose herd. When he was eighteen he was sent to work in the silver mines in Hartz, but still the peasant collected his wages. So when the government offered a subsidy for going to Australia or America, Christian was one of the first to accept it. He arrived in Australia seven years before me, and although working conditions were much superior and food more plentiful than in Hanover, the feeling of being a servant obliged to do what the master ordered, irked. As soon as gold was discovered Christian went to the diggings and soon had over £100 in his pocket. That's when the temptation to live like a lord came, to have all those things he dreamt of, but never had. Let's taste this life, he thought. Back to Melbourne then, but his money didn't last very long. After a fortnight he was back on the diggings. Lady Fortune was kind again: in a short time Christian had another lucky strike. And again for a short time he lived like a lord until all the money went. Since then he had eked out a living and no more, but was confident that his luck would change.

So on 28 December we left Forest Creek by the same road by which I had arrived a little while earlier with my companions. When we left the valley and found ourselves in some open spaces among the forest, we saw smoke rising, indicating bushfires. In fact, as we kept going, miles of bush around us smoked, and flames licked the sides of the wide road along which we moved. It was not a fire to compare with the American savannah fires which drive all bird and animal populations away, it was rather a slow, smoky fire which consumed the growth around. Not all Australian bushfires are so gentle. Later I experienced some which could easily be compared to the American fires.

We spent the night near a stream below Columbine and left there at daybreak to stop for rest at Kyneton, where unfortunately my companions found some gin. After Kyneton we continued to Karlsruhe and instead of following the Melbourne road, we turned left into a large defile and carried on for a few miles. We pulled up for a midday meal and a rest and then continued along the same defile.

Here was thick grass which, growing through old, dead layers of previous years' grass, formed a thick soft carpet. The grass is entirely different from that found on large European meadows. Here not a flower shows, in contrast to the European meadows which are half grass and half flowers. And in colour the grass resembles wheat which has turned to straw, rather than the juicy green to which I have been accustomed. In spite of this it is very good fodder and cattle and sheep do very well on it.

The low hills on both sides of the defile were covered with age-old forest, so tangled as to give assurance that no human foot has ever entered it. Pools of putrid, stagnant water lay everywhere. The silence was broken only by the screeching of cockatoos and the laughter of a grey bird called the Laughing Jack.

So we travelled for a day and a half without meeting a soul until we came to a station. Until recent times only wool and tallow were exported to England and the local population was too small to consume much meat out of the large and rapidly expanding herds. Prices were accordingly low and a fully grown ram fetched three shillings. Now the price has gone up to twenty shillings because several hundred thousand miners consume large quantities of meat and the town population is also growing every year. Settlers have made fortunes on the sale of meat since the mines were opened. As a matter of fact they were doing very well even before, because the mild climate, ample feed and lack of predatory animals make animal husbandry a very easy and simple affair. The only damage (not very great) is caused by wild dogs. These dogs resemble domesticated dogs except that all are yellowish in colour and have shaggy tails. Their only special characteristic is that they do not bark, but howl.

We arrived soon after in Kilmore, where we had to shoe our horse because he had lost one of the shoes on the rough road. The blacksmith charged ten shillings for each shoe. Outside Kilmore we fell in with ten English and Irish miners who had lived in Australia for many years. According to their stories a few of them had been very successful in finding gold and just as successful in quickly spending their money. We travelled with them through the forest to meet the first day of 1853, on the road.

How very characteristic are the Australian forests with their various trees all so different from our oaks and pines and beeches. Here the trees stand far apart, with the leaves at an angle which gives little shade. Here one sees the Box and Water tree, Camphor and Pepper trees, Iron Bark and Light Wood, a tree called Honey Spelt and the Gum tree. All this stands in still, hot air—I was too hot to engage in botanical excursions—and in such a deserted landscape that in the twenty-four hours since we left Kilmore we saw not a single settlement nor one traveller.

Finally we came to a river, not wide but deep. A ferry takes people across. This road leads to the town of Goulburn on the other side of the river and on to Sydney, so that the ferry owner makes a lot of money.[1] The town of Goulburn on the far side of the river is rather a joke. One house which is a hotel, a shed with a large sign "General Store", another shed with parts of the roof and walls missing, two tents near a muddy dam in which two geese swim and watch a pig wallow in the mud on the shore. And this is called a town. Who knows, maybe one day it will be a large town full of life. Rome wasn't built in one day either.

We crossed the town and travelled for two more miles till we came to some grassland on which our horse could feed. Here we stopped for the night and the following day, which was Sunday. Our Irish companions made their camp near us. So we spent Sunday 2 January resting near the river, swimming and fishing.

We started before sunrise the following morning in an attempt to cover as much ground as possible before it got too hot. We travelled in this fashion for two days. On the third day we met a police patrol searching for bushrangers who had attacked a settlement, ransacked the house and killed two of the inhabitants.

Soon after, I lost sight of Gayer and decided to stop for the evening and wait for him. He didn't show up that evening or night, and worried by what we'd heard from the police patrol we decided to search for him. As I was a stranger to the Australian bush and likely to get lost myself, I was left behind with our gear. After a while two miners passed me and on my enquiry told me they had seen such a man with a white dog in company with some Irishmen camped at Broken River. When by companion returned we put the horse between the shafts and went back, where we found Wilhelm Gayer with our erstwhile Irish fellow travellers. Wilhelm explained that he had got lost, was afraid to travel alone because of the bushrangers, and decided to continue with the crowd.

Here also I saw for the first time native women and their children, called *piccaninni*. One of their men who arrived soon after, on our request for fish just dived into the river and soon came out with a tasty foot-long fish.

After we left Broken River we were joined by three wagons carrying goods for the mines at Ovens. The owner, originally born in Memel, was a sea captain. After a sojourn in India and then in California, he became a merchant in Australia. Unused to travelling on solid ground, he leaned a little too far out of the cart. When it struck a large stone he fell out and broke an arm, which I helped to set.

We continued our travels and when we pulled up in the evening some natives came up with possums they had caught. I was surprised to see that they did not skin the carcase but pulled all the hair out. Apparently they bake them in their skins. As we were running out of meat and the moon was out, we shot a few possums for breakfast. The meat is brown and resembles a hare, but is not very tasty.

Eventually we reached the Ovens River and the few houses which were the nucleus of a city. The river, no bigger then the Goulburn or the Broken, feeds into the River Murray, which flows through three colonies: New South Wales, Victoria and Adelaide. Here I also saw for the first time a snake six feet long, and grey in colour. It is not as dangerous as the Black Snake, called the Death Snake, whose bite is fatal within a very short time.

It was Sunday again, so we remained where we were and as we had no meat we shot a few cockatoos and had broth and boiled bird. The white cockatoo is most prevalent, but occasionally there is a black one with a pink crest, a very rare variety.

We were camped near a station and talked with the shepherds, who told us that some years previously the natives had attacked the station and killed the son of the owner. They had also attacked a few more neighbouring settlements. The revenge for the few white men killed was gruesome. The settlers formed a company of about fifty well armed mounted men and for a few weeks hunted all natives in the vicinity. Several hundred natives were killed. I have also heard that some settlers give the visiting natives food, after eating which the natives die. How far this is true I don't know, but it is a fact that wherever a settlement is established the number of natives diminishes rapidly. In Van Diemen's Land only about ten or fifteen natives remain alive, thus leaving this beautiful island for the sole use of their English brothers.

We are only sixteen miles away from the diggings at Ovens. The road rises into hills mounting higher and higher. At one place I could see over the forest into the country. On the horizon are still higher formations, also densely covered with forest. Trees grow everywhere over undiscovered and unmeasured land. The old ones weary with age fall over on to other trees and form unpassable obstacles. The younger generation of trees shoot up. In their shade glide silently, like ghosts, groups of natives unclothed and untroubled by any knowledge except that necessary for survival. Who is to say that they are not happier than we shall ever be with all our civilization? The mild climate, rivers full of fish, the bush filled with kangaroos and possum supply their needs, while their unerring instinct surpasses all sciences.

We shall be at the diggings in a few hours and the countryside is becoming ever more hilly. We pull up at a gully where there is a stream and plenty of grass for the horse. The ten Irishmen pull up with us. We don't bother putting up the tent, because it's early enough to walk to the diggings and investigate where we shall peg a claim before selecting a place for a camp. We walk down a gully and see a thirty-feet-high waterfall. We are unhappy to see quantities of water going down into the valley where we are supposed to work, because it means that the soil in the valley, locked in the hills, will be continually overflowing and the shafts will have to be shored up against underground streams. This proved to be correct, for all shafts although not very deep had to be lined with planks. Some miners' companies use buckets to empty the water, others use all sorts of pumps, but in either case it means continuous work to get rid of the water. This is impossible for two or three miners, so we immediately arranged with our Irish companions to form one company for the duration of our stay.

We marked a claim and returned to the camp to cut trees for suitable planks and stakes. It took two days to collect sufficient planks and stakes and a day to bring them to the site. On the fourth day we began digging. The work went well and by the evening water had begun to seep into the shaft. Early next day we arranged for three men to dig and throw the soil out, two men to move the soil away from the vicinity of the shaft, one man to fix planks and stakes and six men to form the bucket brigade. By the evening we were nine feet down and near the bottom. Next morning water was chest deep in the shaft and some stakes were bending and twisting with the stream. It was obvious that it was necessary to bottom as quickly as possible and collect the washing stuff before a cave-in.

We worked without a lunch break and at the end of the week, when we divided our gains, we obtained eighteen shillings each, just about what it costs to feed a man for a week. Some poor devils don't even get that much for their hard week's work. I think it would be a good idea if some of the speculators operating on stock exchanges and who claim that the price of gold is bound to come down because of the ease with which it is found were to come to the diggings and work for at least a day. They would soon give up this foolish idea that digging for gold is easy.

We dug another similar shaft the following week with

the same back-breaking toil, but we barely got traces of gold. We also found the company of our Irish partners rather difficult to take. Drinking, swearing and fist-fights amongst them, to be forgotten in a few minutes, were surprising and inexplicable to us. We tried to sink another shaft, but when this was also proven to be barren we decided to follow the lead of some other miners and quit the Ovens diggings.

We heard a whisper of a new find at Buffalo and decided to start in that direction, hoping to hear more about it on the road. We joined a company of Scotsmen proceeding to Buffalo. We travelled along less hilly terrain, full of rich grasslands well watered by the River Ovens. There were tall trees, plentiful shrubs and grass tall enough to hide a man, but like elsewhere in Australia not a flower amongst the grass.

At sunset we pulled up near the house of an overseer guarding cattle for a settler—if you could call this leaning shed built from split planks a house. In the main room a corner serves as a fireplace, with a kettle hanging from a chain over it. One shelf with three plates, a table made from stakes, and a bench, form the whole of the furniture. The other room has a few bags filled with straw or hay for beds. No windows, just square openings which can be secured with shutters in case of need. No garden or vegetable plot, just two graves outside, a few chooks, a turkey and a pig with her litter. We bought some milk at one shilling a quart and beef at one shilling a pound.

The Buffalo mountain range was only thirteen miles distant, but we decided not to go there when we saw two wagons moving the contents of the only shop on the diggings out of there. The news is that there is no gold but many bushrangers in the vicinity, so it's no use going.

I saw an amazing sight there. The Scottish teamster had backed his oxen to a large fallen log and he began beating them with a cudgel as hard as he could. He was not angry, nor did the oxen cause any damage or misbehave, yet he went on punishing them. When asked for an explanation of his behaviour he told me that Australian oxen are half-wild, bred as they are in the bush, and are likely to kill or maim a man because they resent having been caught, deprived of their freedom and made to take orders. The only way to control them is to show them who is the master, hence the periodic hidings.

On our next rest I saw for the first time in Australia a plain clear of bush. In the distance to the left we could see Mount Buffalo. We spent our next night at Kings River. The following day we travelled twenty-six miles in exhausting heat until, after crossing the Fifty Mile Creek, we pulled up at Broken River. We were uncertain where to go, for we had heard of new diggings near Devils River, but nothing definite. As we were at the crossroads, we could go to Forest Creek or take the right fork of the road to Bendigo. We had to decide. The decision was in favour of Forest Creek and we spent the night at Seven Mile Creek. A day later we pulled up at Julia Creek, where we found numerous groups of natives. Soon after we crossed the Goulburn River and camped for the night two miles further on.

After we passed Kilmore, we went towards Karlsruhe. Thus we returned after a few weeks to the spot where we'd camped on the way to Ovens. Plenty of green grass and reasonably fresh water in the pools made it a good camping spot. Just as I was falling asleep, Wilhelm drew my attention to a noise like faraway thunder. "It's a bushfire", said Wilhelm, "but still very far away." About midnight the dogs woke us with their barking. All we could hear was the breaking of twigs in the bush, so we concluded that kangaroos were running away from the fire. We didn't worry about our own safety, because we were shielded from the approaching fire by a wide road and some large waterholes. As we were readying the camp in the morning, baking bread and washing our linen, we saw clouds of smoke as far as the eye could see and heard the growing thunder of the fire and the falling trees. The fire reached us an hour after noon. Only then did I get the idea of its size and from how far away it had come.

I watched the gruesome sight—a sea of fire rolls forward, dry and green grass take alight, tongues of flame lick large trees in passing, marking them with a black imprint, gnawing at dry trunks, felling them to the ground with a dull thud. The low bushes just disappear. Woe to the man or beast caught in it. There is no deliverance, because the high grass, never cropped by herds, burns like tinder. This fire I watched rolled on a front well over ten miles wide.

We were indeed fortunate that the wide road and the water had stopped the fire just a hundred paces from us. The fire burnt all the following night with the thunder of exploding, flaming trees and the noise of falling trunks. A few trees which burnt standing up looked like lighted obelisks in the night. When we took to the road again it took a few days before we arrived in Forest Creek to look up my old companions. Two had left for the diggings at Mount Couron[2] and one had passed away after a long period of weakness and fever.

7

Goldfield incidents and stirrings of revolt

Because I was in debt to Wilhelm for my keep, I had to remain with him until his prophecy that we should find gold came true. I believed in his conviction but felt that it was taking an unusually long time to come to fruition. My patience was wearing thin in spite of repeated lifelong lessons that patience is a great virtue.

As I mentioned before, all diggings have extensions in various directions. Sometimes such arms extend for miles and each has a name, just like roads in a big city. The names stem usually from some conspicuous feature or the name or characteristic of the miner who first dug there. And so at Forest Creek a major district was Pennyweight Gully, because gold was found there in small quantities; Gold Gully where gold was plentiful; Adelaide Hill, Montgomery Hill, Campbells Creek and many others. In Bendigo we came across Dead Horse Gully, Peg Leg Gully, Napoleon's Creek and Beelzebub Creek. On the Alma diggings where we were the first to discover a new arm of the gold-bearing strata, the place was called Polish Flats.

Wilhelm Gayer decided to try his luck at Adelaide Hill. In the first week we dug three empty shafts, in the second week we were more fortunate, and in the third week I paid all my debts to Wilhelm and had a few ounces of gold in my pocket.

The gold washed every day is dried, then dust and other light impurities are blown off, bits of iron are removed with a magnet and the rest is minded by one of the miners until Saturday, when it is all shared among members of the company, according to agreement.

We kept hearing from all sides about robberies and murders. Most of those happened on the Ballarat road and

Gold Point, Mt. Alexander. This engraving by Thomas Ham was made from a drawing by David Tulloch and was originally published in *Five Views of the Gold Fields of Mt. Alexander and Ballarat in the Colony of Victoria*, 1852.

some near Geelong, but even at Forest Creek things were unpleasant. Some of the robber fraternity pretended to be miners, but took more interest in who had a shaft yielding a lot of gold than in digging a shaft themselves. Two of our neighbours, young well-built Dutchmen, were warned that someone was watching their shaft. They made arrangements for us to come to their aid should they call out in the middle of the night. We heard a commotion next night and went to the Dutch tent. They had one bushranger tied up with ropes, his head split open with a pickhandle and the other bushranger was dead. The Dutch boys told us that they had spent the night behind their beds instead of on them and when strangers armed with knives entered, surprised them with a sudden attack. The dead man was unceremoniously buried and the other handed over to the police commissioner.

Easter Sunday came and I spent it with my countrymen, who worked on the Campbells Creek diggings. I was told of two cases of miners shooting thieves. One happened during the night when a miner saw a thief cutting open a side of his tent, and the other when a miner, passing his shaft at night, found a stranger there picking up the bottoming earth. It sometimes happens that dead men who may have fallen in accidentally in the night are found in the shafts.

For a few weeks we had a good run and each of us had gold put aside. We heard some encouraging reports about the diggings at Mount Couron and decided to spend the winter there. Mount Couron is eighty miles further inland from Forest Creek on the banks of the Loddon River, which actually becomes a river only in winter. Assisted by heavy rains the Loddon overflows its banks and floods the countryside, cutting off all communication with Melbourne. The merchants exploit the situation to raise the already high prices. Consequently those miners who can afford it, buy flour, coffee, tobacco and other provisions in cheaper places before they travel to Mount Couron. We discussed the problem one evening over the fire and decided that Wilhelm and Christian would go to Melbourne to buy supplies from the cheapest source, whilst I was to remain on the diggings to wash the soil remaining from our last shaft.

My companions returned after about five days and we made a start for Mount Couron. We had too much gear and supplies to make a single trip of it, so we left Mrs Gayer with some of our goods at Forest Creek. We had to stop at Barkers Creek because the rain was too heavy to make travel safe. Next day we deposited our gold and money at the police station at Gibron,[1] because the roads were full of bushrangers. The next night was more comfortable because it was not raining, and I found to my suprise that we had already crossed the Loddon some little time previously. According to Wilhelm, the dry defile we crossed was the Loddon River in winter. We shot a few parrots and had broth and bird meat. Both quite edible, but it was with sorrow that I plucked the beautiful sky blue, scarlet and green feathers mixed with black and lemon coloured ones, with pale grey and light pink ones.

We arrived at Mount Couron and inspected the diggings, where only a small group of men worked. None of them looked happy and one could not hear any shooting of an evening, which was usually a warning to the bushrangers to keep away. Apparently no one was rich enough to worry about bushrangers or to spend money on gunpowder. Food was terribly expensive. The locality didn't make a good impression on us.

Mr Read, the mining commissioner and an acquaintance of mine, was in charge of the district so I decided to ask his advice about remaining there. The advice sounded more or less as follows:

> You can find gold here as well as anywhere in Australia, but here one has to look for it more intensively. You may have to try forty or more shafts before you find anything. My advice is, look for it where it is easier to find. This is an accursed place. Drinking water has to be carted from five miles away, even though in winter the creeks overflow and we are completely cut off for weeks at a time from everything and everyone. My personal view is that one ought to escape from here and if possible leave Australia. This place is fit only for Negroes, Chinamen, deserting seamen and

ex-convicts. It's not a place for a decent man. He would have to work himself to death and grow wild. As soon as I save enough I will clear out from here.

I don't think Australia agrees very much with Mr Read. If he is willing to quit his position as commissioner on £300 per year, how then should a poor miner react? I repeated what I heard to my companions and we decided to go to Bendigo. We sold some of our flour at a very high price, because we had more than necessary to last us all winter, and left Mount Couron.

Now it rained every day. Fortunately the road was sandy and soaked up all the water. Walking was not difficult but our shirts would soak up the rain as well and we had to pull up early every evening to make large fires to dry our clothing before the night's rest. Wet earth doesn't make a healthy or comfortable bed. Sometimes we spent all night sitting down near the fire in a reasonably dry spot to catch short naps. We crossed the Loddon River, but this time not dry shod because the rains changed the dry river bed to a sizeable stream which in a few more weeks would become a wide river if the rain continued. Two days later we approached the Bendigo goldfields, thirteen miles wide and employing 20,000 miners. Everywhere signs of work on a large scale: one hears the explosions coming from the hills where gunpowder is used to loosen up the hard soil. Everywhere one can see the white cloth chimneys over the shafts, a sure sign that they are deep, and big heaps of earth for washing are piled up near the shafts, indicating that the place gives rich yields.

We put up a tent and Wilhelm went back to Forest Creek for his wife and the balance of the gear while Christian and I marked out a claim and began to dig. At a depth of seven feet we came across a stratum of oxidized iron mixed with quartz. We had to use steel wedges to loosen this stuff, a back-breaking job and dangerous because often the wedge jumps out from its place and hits the miner.

When Wilhelm came back, he brought with him his wife, the rest of the gear, and a lamed horse. It took a few days to cure the horse, and then we arranged that I would leave my two companions to dig our claim while I took on the job of carrying washing stuff to the water for other miners. I made £4.10 per day with one horse and a two-wheeled dray, so that at the end of three weeks our profit from this undertaking was £102. This may be difficult to believe, but those who know Australian conditions would not doubt it. If such profits were to go on over a number of years, one could make a handsome nest egg. Unfortunately the harvest lasts only as long as the claims located on the hills are being worked and it's far to the water supply, which in such hilly country applies even in winter. After spending money on feed for the horse (£2 per bushel) the rest was divided between the three of us to pay for our needs, because during all that time my companions dug two shafts and found no gold.

At this time I became involved in smuggling grog. I was hired by two Prussians to carry their goods to Inglehook,[2] loaded my dray in good faith and proceeded with my customers along the road when my companions told me to get off the road and drive through the forest. When I didn't want to do that they told me that a police patrol was soon due, that there were contraband goods on the cart, and that both the cart and the horse would be confiscated if we were caught. It was too late to take any other action so I had to turn aside into the bush and reach Inglenook that way.

My next longer trip was to MacIvor diggings, shifting a group of Scottish diggers and their gear there. After my previous experience, I loaded all the goods on the dray myself, to find out that this time there was no need for anxiety. My customers were bona fide miners and carried no spirits with them.

After returning to Bendigo I found the Degenhardt brothers there and carried their washing dirt for them. They had two good shafts and used to wash out a few ounces of gold daily. Here I saw also for the first time a curious bird the size of a crane, which stood every day near Carl Degenhardt and watched him work while he washed the dirt. At first I thought that the bird was tame and belonged to Carl, but I soon found that he just visited Carl out of some sympathy, or liked the particular spot where Carl worked. The bird is called the Native Companion.

Several natives often visited our neighbourhood asking for rice, sugar and tobacco. Once when they were outside the tent talking, one of the native women suddenly turned her head to the side and seemed to be listening to something. After a while she jumped to a solid tree nearby and with a tomahawk split the bark and pulled out a white grub about four inches long which she ate on the spot. Asked how she knew the worm was there, she answered surprised:

"But I heard it. It was only a few steps away."

The natives have a wonderful sense of smell, hearing and sight which no white man will ever achieve. They have also a sense of direction which allows them to travel a few hundred miles by the shortest route to their goal, through monotonous bush without any landmarks.

After three months in Bendigo, my companions did not find enough gold to buy a pipeful of tobacco. If not for our horse, we would have had to draw on our capital brought from Forest Creek. Poor Jack gave up eventually and could pull the dray no longer. We had to sell him and I helped with the digging. However Lady Luck, who smiled at others, didn't favour us. We even tried to search for gold in the bush six miles away. In spite of the conviction that where others work and prosper we should also be able to make a go of it by sheer hard work, we began to doubt if the place was meant for us.

Eventually we decided to shift to Myers Flat, a branch of the Bendigo fields about five miles distant. Even here we sank eight shafts without results, but eventually our luck turned.

Once, walking by, I noticed two tattered tents standing on their own. The four Irishmen living in the tents worked two days a week in some holes in the creek which looked poor and long neglected. The rest of the time they drank and fought. Seeing them thus day after day, I concluded that they wouldn't fight continually unless they were drunk. And an Irishman doesn't get drunk on one glass. A three pint bottle of brandy then cost £1. To fight and drink every day required a lot of gold.

Thus reasoning, I approached one of them and asked how far their claim extended. The reception was hostile and every attempt was made to make me quit, but I quoted the mining law to the Irish and eventually forced them to indicate the limit of their claim. Then I marked my claim and started to work it the following day. It was a shallow shaft and next day I came to the bottom. My line of reasoning turned out to be correct and I found a fair amount of gold. I told my companions about my find and they moved next to me. Their claim turned out to be not as rich as mine, but not bad. However others who followed our lead and tried lower down the line had no luck. The conclusion was that there was no more gold in this direction, but a few days later quite a number of miners turned up to try their luck still further down and struck it rich. When we first saw them there we laughed, but later felt sorry we hadn't tried there ourselves. Such is the luck of the mining game, for one cannot foresee where to look for gold.

I was so disappointed with myself for my lack of foresight that I shifted to the Sydney Flats about two miles away and uselessly dug three shafts about nine to ten feet deep. Near one shaft was the grave of a native chief. In an oval of young gum trees was a place completely clear of grass, surrounded by a shallow trench. In the middle was the grave, an oblong hole in which was the dead person, not buried in earth but covered with bark. Apparently such a grave is made only for very special personages, because although I criss-crossed the bush for three and a half years and came across a number of natives, I never saw a similar grave. It would have been an excellent find for an archaeologist but I was too busy sinking my shaft to worry overmuch about scientific problems.

At the Sydney Flats I met a Polish blacksmith from Bydgoszcz by the name of Pieczko. Some years ago, when he was in Hamburg, he had signed on as a blacksmith on a ship going to Australia. Here he gave up blacksmithing and became a miner. He told me in secret that somewhere in the bush is a very rich place, where four miners live in two tents. They—said Pieczko—had worked there for three months already and brought their gold in a bottle to the commissioner. Two of them came every Saturday, bought their supplies and quietly disappeared. Once Pieczko followed them, but he was spotted and the miners sat down on a log and waited until dark. Pieczko had difficulty getting back to camp, but as he knew the direction in which the miners went, he decided to follow them again. I was allowed to take only one mate and join Pieczko in the new claim after we found the location. I told the story to Wilhelm and we both waited impatiently for next Saturday to have the mystery solved. We found that our new friend managed to follow the miners up to some dense bushes, where they disappeared. He circled the bushes but did not see them leave. Although he didn't want to risk being discovered and possibly killed in the dense bush, he was now certain that he knew the location of the place.

We arranged to start the following morning, Pieczko with his own companion and I with Wilhelm. After a few hours we came to the densely overgrown patch of bush, which was our goal. In spite of a careful search we could find no trace of the diggings and were very disappointed indeed, when suddenly we saw two miners coming our way. They greeted us in a friendly fashion and asked what we were looking for. We told them outright that we'd heard there was a rich patch in the vicinity and we were looking for it. The miners laughed goodheartedly and led us to their camping ground and some shallow shafts. The full story came without much prompting. One of the four miners was sick and not well enough to move or be moved anywhere. His companions were getting a trace of gold, about a pennyweight each per day, and were living on that while waiting for the sick man to recover.

This was the end of our adventure looking for the "mother lode" of gold, the dream of every miner in Australia. It is a popular belief that only a small part of the original "mother lode" comes to the surface in small nuggets while the core remains deep inside. Naturally every miner dreams that he may be the chosen one who will discover this untold wealth.

I returned to dig a shaft in a new environment. Three of the nearest claims belonged to Maoris, the fourth to Chinese. There were then between 12,000 and 14,000 Chinese in Australia. They frequent mainly diggings with shallow shafts. Those "children of the sun" don't seem to like the darkness of underground workings. Most often however they collect the sand already washed out from the miners and go through it again, finding a small residue only, but without the back-breaking job of digging. They sit comfortably day in and day out with their small cradles near a stream and earn enough for their modest needs. Their staple food is rice and tea, rarely meat. They show amazing perseverance and patience.

There is a continuous inflow of Chinese to Australia, but men only. The colonial government has a problem on its hand for now that the monthly tax on miners has been cancelled the Chinese have ceased to be a source of government income. The merchants also don't seem to be able to sell them anything and thus increase their turnover and profit. Their continuous inflow threatens that they may become more numerous than the European population. Recently the legislature debated a project to send newly-arriving Chinese north towards Cape York. Now that there is a revolution in Canton, the inflow may be tempered, or the British Empire may possibly use them to populate New Victoria (in Antarctica), discovered by Captain Ross during his explorations.

The Chinese are small in stature, with small eyes and long plaits of hair made even longer by a piece of string with a tassel tied at the end of it. They are very funny to watch when they walk overland, for they usually travel in large groups of a hundred or so, one behind the other in a long line like wild geese. They don't walk normally, but take short steps and appear to be running very slowly. Each one carries a long pole over his shoulder with baskets of victuals hanging at both ends.

The New Zealand natives are entirely different. Tall and strongly built with well developed features, they could be called a handsome race but for tattoos on their faces and their black, wildly lustrous eyes which remind one that in spite of the missionaries and colonists settled amongst them in their country they are still counted among the cannibals. Their wild habits have been described by Tasman, Cook, d'Entrecasteaux, Duperré, Laplace and others. Revenge is said to be their duty and devouring their enemies their pleasure. I read recently in a Melbourne paper that two Maoris had been caught in New Zealand selling human flesh as pork.

My Maori neighbours dressed and worked like all the other miners. Of the six I knew, I knew the names of only three: Tatika, Jehova, Morka. They were very kind and courteous to me, probably because of my grey hair, for I understand they have a lot of respect for age. They always put my billy on the fire at lunch time with their billies, so that when I came out of the shaft I had hot water ready. Two of them had some English and kept telling me about the beauty of their country. They seemed very friendly and pleasant to all around until the following incident occurred.

When their shaft petered out they moved to Sailors Gully and had there a good shaft. A couple of Englishmen wanted to take it away from them and one came early one day to start digging while the other was at the winch. One of the Maoris arrived, and seeing a stranger in his shaft, picked up a large stone to drop on the intruder. He was restrained by the man at the winch. The other Englishman, hearing the commotion came out of the hole to help his companion. Realizing what was taking place, the Maori took out a revolver and calmly shot both the intruders dead.

Shooting and murders seem to be quite common. Just when this incident took place, we received news about an attack by bushrangers on a gold escort going to Melbourne. The bushrangers had spies who advised them when the gold escort was to leave MacIvor. They surprised the escort in the bush, where they blocked the road by felling trees

"The claim disputed". This lithograph by S.T. Gill was originally published in *Sketches of the Victorian Gold Diggings and Diggers As They Are*, 1852.

across it. During the battle between the immobilized escort and the bushrangers, six members of the escort and two bushrangers were killed. The bushrangers left the field victorious, with 3,000 ounces of gold. I heard later that an expedition of about fifty settlers and merchants went after the bushrangers,[3] surrounded them and caught them, recovering part of the booty.

Meanwhile I dug three shafts in search of gold. I had one shaft near a shaft already thirty feet deep which was being sunk by some English miners. I calculated that I could save a lot of useless labour if I were to find out how successful they were. I asked one of my neighbours, who had just returned from testing a bucketful of earth at the washing place, about the result of the test. The report I received was very encouraging so I went on energetically digging my shaft. During a break a compatriot of mine passing by stopped for a chat. My English neighbour was listening in and came up to me later asking in what language I was conversing. "My native Polish", I replied. Apparently national animosities exist as strongly on the fields as anywhere else, because my neighbour explained with a great deal of embarrassment that his test hadn't shown any trace of gold and that he had misled me because he thought I was a German.

The result was that I changed location again and found myself surrounded by the following: a man from Holstein with a partner from Vienna, two Bavarian partners, two Americans from Honduras and an Irishman with a Dutch sailor mate who, after being caught by cannibals in the Java Sea, being sold to a Japanese pirate as a slave and then resold to a Chinese, managed to run away and eventually reach Sydney. His bad luck seemed to keep up with him even on the goldfields, because although we all had good returns from our shafts, he could not find a trace of gold and finally decided to go to Bendigo where he had a prospect of a job as a tentmaker. We passed the hat around and made a collection for him to help him on his way.

About this time the European newspapers announced the death of Wellington.[4] During the following few days my English acquaintances continually bothered me with the question: Do you know that the Duke of Wellington is dead? No doubt Wellington served his country well, enhanced the fame of the British army and has been outstanding among British generals, but all this doesn't concern me sufficiently to be exposed for days to the repetitious listening to the sentence: Duke of Wellington is dead—and to be obliged to reply a hundredfold: "Yes".

The English have good reason to sigh, pull long faces and pay homage to the warrior. And so they sigh now, after they have lost him; but no so long ago, when he was a Tory cabinet member they used to break the windows in his home. The victor of Waterloo was obliged then to fit shutters to his windows to protect himself from the ardent demonstrations of his compatriots. And so I was quite happy when an American visiting my shaft cut in on the continuous homage offered by my English neighbours. "What's the matter with you? You keep pestering everybody with Wellington. If you haven't anybody better in your country you are very poor indeed."

"Oh you have plenty of famous people", replied the Englishman, "and the foremost are General Jackson and Jesus Christ."

This is a standing joke at the Americans, because once two were overheard in a discussion over a bottle about famous Americans. One of them, not too well read in history, knew only about General Jackson and therefore claimed that he was the greatest American. The second, somewhat better versed in the affairs of his country exclaimed: "What, only Jackson?"

"No", replied the other, but short of other names or possibly too full of gin, corrected himself: "General Jackson and Jesus Christ."

The English pinprick the Americans about this at every possible opportunity.

This time my American friend calmly listened to the allusion and just as calmly replied: "You yourselves have fabricated this foolishness and are attributing it to us, but regarding great men, I expect you'll agree that your Wellington is no patch on our Washington's boot."

Judging by the clenched fists and raised brows I doubted whether the Englishman was prepared to accept this statement peacefully, so I expected a physical protest. But luckily the intervention of some Germans who sided with the American prevented this.

Some representatives of the German miners had been busy influencing the various nationalities in Australia to an uprising against the government. Meetings were held every week on all the diggings, with speakers using fiery expressions to foster unfriendly feelings against the government.[5] Lacking arguments, they insulted the Queen, trying to deride her and calling the government in England a petticoat government. They cited the civil list and the huge amounts of money spent on the Queen, her husband and her family, as well as the pensions paid to the progeny of the mistresses of previous kings. Comparing those expenses with the expenses of the United States, Mr Bruun, one of the main leaders, claimed that the British court alone cost England more than the President, the whole civil service, the army and navy in the United States.

"And where do those large sums to meet the large and unnecessary expenses come from?" he called loudly.

> From the bloodthirsty exploitation of the work of millions of people all over the world. To it contribute the hunger-ridden Irish, the depressed Scots, poor workers crippled by hard work in English factories, cheated Turks, forcibly enslaved Chinese, and the colonists in many countries and islands, amongst them us Australians. The government sells forests which do not belong to it at exorbitant prices, exacts larger taxes for permission to dig in ground which does not belong to it, and shows no concern whether the thousands of us find enough gold to make a meagre living. To keep order in the colonies some taxes are necessary, but let us have a just system: those who have pay, not those who are killing themselves in fruitless work. Let the tax be placed on gold exported to Europe and not on the diggers. Alternatively let the government cease sending gold to England. As it is they charge us enough for imported goods so let the gold be used for making roads and for improving and enlarging the colony. But who can take this up with the councils in Melbourne?

There sit those who were deported for crimes, while we miners, being the bulk of Australian population, are not allowed to be represented. Is this justice?

Each sentence was greeted by a rising "Hurrah" from the gathered miners. A committee was elected which established contact with committees on other goldfields and began to act. In the first week the Bendigo committee announced the following decree:

1) Miners should not pay the monthly licence fee. Whoever shall be seen paying such a fee is warned that his tent will be set on fire.
2) Armed miners from the vicinity are to meet next Saturday in Bendigo, march together to the commissioner and present demands to abolish the licence fee and to select a deputation to go to Melbourne. Similar action is to take place on other fields.
3) Every miner is to wear a red ribbon in his hat as a sign that he supports the action.

The miners stopped paying taxes, in a few places the police inspecting the licences were manhandled, and on the chosen day a few thousand diggers gathered in the Bendigo with firearms in their belts. They were divided into their national groups designated by their national flags. The Chinese followed a long pole with small bells hanging on it; a star-studded wreath preceded the American group. Music and continuous firing of firearms warned the chief commissioner that the miners were approaching. They stopped before the encampment where the police and public servants had their tents, and twenty selected miners stepped forward to enter the tent of Commissioner Panton.[6] This wise and polished man received the armed deputation with a smiling face and a polite inquiry:

"What can I do for you gentlemen?"

After listening to their demands, he replied slowly:

You know gentlemen, that I have no power to reduce or remove the taxes, nor can I select a deputation from your body. If you don't want to pay taxes, I can't force you because I have not sufficent strength to do so. I'll make representation on your behalf to the government. You send someone who

Great Meeting of Gold Diggers Decr. 15th 1851. This engraving by Thomas Ham was made from a drawing by David Tulloch and will know how to support your demands and let us wait for the results. In any case, I warn you gentlemen that the government could send troops if it decides to retain the previous regulations.

"Let them send troops. Most likely they'll never see those soldiers in Melbourne again. We know the strength of the army", replied the deputation.

The commissioner sent a report to the government, a few selected miners from each diggings were sent to Melbourne and 200,000 miners[7] stood grimly by their resolve to cease paying the licence fee. The conspirators may well have had other plans, because I heard that the first demands were to open the way for further action.

I did not attend the meeting and did not wear the red ribbon in my hat, but neither did I pay for the licence —why should my brother diggers burn my tent? I worked solidly in my shaft, watching the behaviour of others. All the time young, hot-blooded men arrived and made speeches. Eventually a few approached me, with an American in the lead. He greeted me normally and then asked in a conversational tone:

"Do you know what's happening in Australia?"

"I don't know, it is only a few weeks since I began work on this field."

That's time enough to know, because the miners everywhere have decided on similar action. We believe that the fate of Australia will be soon decided and that she will take her due place among the nations of the world and become just as happy and prosperous as the United States. Everything is ready for the struggle. We are approaching you as an old experienced soldier, asking that you should at the right moment organize the miners on military lines and lead us, if the need for it arrives.

I replied:

This is a great honour you pay me, Gentlemen, and with equal candour I am bound to answer that I will not organize nor lead you, because I cannot be ungrateful and I owe a debt to the British government. When we exiles were barred from most countries the British government alone gave us

[illegible] 1854. This lithograph by S.T. Gill was originally published in *Sketches in Victoria*, 1860.

shelter. This I will never forget. Did you ever consider what you are trying to undertake? Have you considered the means which Australia has and are they sufficient to maintain independence without help from England? No, the time has not yet arrived for Australia to imitate America. You could topple the present government in the colony and, as you say, be free. But how long would this freedom last? Any flotilla arriving in Port Philip Bay could land a few thousand soldiers and a battery of guns. Could you withstand even such a weak force without arms and sufficient trained men? Do you believe that all those who cheer during the speeches and march to music, playing soldiers, would withstand an assault by a trained army? I assure you no, especially if there was need for long marches and hardship. I know that a leader who would leave the diggings with a thousand men would find himself a month later alone, possibly with an adjutant, as once happened to Washington. If you could finish the whole affair in one battle you might stand a chance, because the initial enthusiasm would keep you all together, but you would not last through a prolonged campaign.

"You are disparaging to us."

No my friend, but I know the difficulties of army life. However you could be an exception. But let me ask if you have received authority to act from all Australians? The miners are only part of the population. There are farmers and city folk, have you asked their opinion? Perhaps they have no wish for the type of liberty you envisage. It would be unfair to force it on them and surely you do not wish to be unfair. You are a spokesman for independence, but at the general gatherings this subject has not been discussed. Only licence fees and representation in the legislature were discussed. Then wait, for perhaps the government will see fit to reduce or abolish the licence tax and to admit your representatives to the council.

"All right, we'll wait. In any case our demands are only the beginning! A wedge for others to come. We shall see what answer we'll get."

The deputies returned from Melbourne with the news that the tax was to be temporarily reduced from £4.10 to £2 per three months. A decision regarding the complete removal of the tax and the admission of miners' representatives to the Legislative Council was postponed.

It seems that Governor La Trobe preferred not to make a decision in this matter as he wished to retire and had already applied to the British government to relieve him.[8] He was awaiting the decision on this retirement, to return to his native Switzerland to enjoy a peaceful life on the pension due to him for his long service in Australia. A governor's pay in Victoria is good, £15,000 per year with a palace in a garden setting two miles from Melbourne near the sea. Mr La Trobe could well save on such a salary and transmit his savings to Switzerland.

This first concession by the government cooled the miners, even those hotheads who were taking part in armed demonstrations. The majority of the miners tried to influence the hotheads to settle down, and claimed that the government did everything possible for the miners. They urged patience. It was obvious that the majority was not initiated by the ringleaders into the real reasons or into the strategy, nor were they aware that the demands just won were meant to be only the beginning. Meanwhile these first concessions deprived the ringleaders of many adherents and forced the temporary suspension of further activity.

8

More mixed luck

The end of the winter season approached and out shaft petered out, although with good profit to all of us. I saw Mrs Gayer, our housekeeper, only at the morning and evening meals because my tent was far away from the Gayers' tent. They left for Bendigo in mid-November and I remained on the diggings a few more weeks in the company of some of my countrymen.

One Sunday I was invited for dinner by the chief commissioner, Mr Panton, and walked to Bendigo.[1] It rained hard so I remained there and spent the night with my countryman K., by then an officer in the mounted police.

I saw Wilhelm Gayer a few more times before he left because we moved, this time to the site of a large diggings to work a claim marked out on a road between New Chum's Gully and Gold Gully. When the gold in those two petered out, the road remained untouched. It was proposed to the commissioner that he allow the miners to peg the road, work the claims and then restore the road. Mr Panton accepted the proposal, but to avoid complications divided it into claims himself and licensed miners to work the claims. To ensure that the road would be restored, each claim owner had to put up a deposit.

We obtained four claims. There were eight of us, but that was not enough people to remove so much surface dirt—at least seven feet. Consequently we admitted two Swiss to our partnership—solid, quiet men. They were hatters by trade, who had gone to America a few years before, and after earning their passage money there, arrived in Australia. They came at the same time as we did from Myers Flat to Bendigo and had a claim near this road. We found no rich deposits under the road but our time and work were well rewarded. Our countryman S., who arrived from France, also joined us. He had a weak chest and was not fit to do hard work, so to give him a chance to earn a living we gave him a job as a cook and paid him £2.10 per week. He was quite busy, because to cook and to bake bread for ten men is a full-time job. He had quite a problem buying provisions, because he didn't know English. We taught him the necessary words, but if he forgot on the way he had to return empty handed. The butcher where he went twice weekly apparently understood his language, formed from three various tongues. He always took the same quantity and it was apparently not difficult to understand when he said: "Give me *dix livres miesa*."[2]

The newspapers brought the first news of conflict between Turkey and Russia. It was amusing to hear the comment of Englishmen about the future course of events, because the English are most conceited: considering themselves the mightiest and most courageous nation in the world, they believe that the fate of all nations depends on their good will or whim. One word of Palmerston is sufficient to decide if there will or will not be a war, if the strongest fortifications and even whole nations are to crumble. I kept hearing boasts about the perfect army organization, the wisdom of ministers, the fitness of generals and the unlimited means for any type of action. This last statement is probably true, so long as the rich colonies are not cut off from the motherland. The custom of buying military rank has caused many grievous losses in the Crimea. The fitness of generals to command is sufficiently well-known and the wisdom of ministers will show in the future. To describe the British feeling of superiority by the English let me cite a conversation I had with an official in Australia:

"Did you read the papers and what do you think?"

"I believe there is going to be a war."

"You are joking. England will not allow it."

"And the burning of a Turkish flotilla at Synopi and skirmishes on the Russian border, where Russia has large armies? Is that for fun?"

"That means nothing. Our fleet on the Black Sea will keep order and we shall instruct the Russian Czar to withdraw his army and that will be the end of it all."

"And if he won't listen?"

"In that case we shall send a division to the Danube. When the red uniforms brighten fields near the Danube he will change his mind. In any case we can order the shelling of Kronstadt and Sebastopol."

"But both of these are supposed to be superbly fortified and very strong."

"What of it? Gibraltar was even stronger."

"And you did not storm it for this reason? I read that you bombarded the conscience of a certain person with sterling and he made it possible for you to occupy the fort."

"Just tales. How long do you think it would take to lay Kronstadt and Sebastopol in ruins?" "I can't tell, not knowing the strength nor the location of the fortifications there, nor the means you could use for the purpose. From what I've heard and read on the subject, it could take a few years."

"What? You'd better say a few months. Even so you'd be mistaken. Our fleet is in the Baltic and Admiral Napier could spend the winter in Petersburg if he so wishes."

"I don't doubt his wishes, but think that he will decide to return to England before winter sets in. It's very cold in Petersburg in winter."

"You are joking because you don't know what we can do. Goodbye."

"Goodbye, but would you like to bet that there will be a war and that you will not take any of those fortifications under two years?"

"I'd stake all I possess on that and I'll be the loser if the first condition of the bet comes true."

"I am not as rich as you and cannot accept the bet for more than ten pounds."

"Right, it's a bet."

A year later my friend admitted he was a little hasty. This conversation is only a sample of the opinion of Englishmen in Australia. I haven't met one who thought that anything could stand up to their might. They changed their mind slightly when the French took the central bastion of Sebastopol in 1855. I would like to hear their views in another fifty years.

It was about this time that I experienced my first whirlwind. At first only dust, leaves and bits of paper are lifted, then the whirl shifts, lifts higher, and grabs everything it meets on the way. An old hat chases a shirt, a lady's veil is caught in a sleeve of a torn coat, the torn half of a vest travels with some other pieces of clothing ripped off a line. This eccentric dance grows into a pillar a few hundred feet high, crosses roads and diggings and vanishes eventually somewhere in the bush.

One of the most uncomfortable moments on the diggings, because of the fear that there could be a long interval of work without results, is when work peters out in a shaft and one doesn't know where to sink his pick. We were in that situation towards the end of December. We took all that was obtainable from our claims and remade the road to its original state. Now we debated what to do. Everyone gave his point of view according to instinct and inspiration, but those things work only for the natives and we couldn't rely on such guidance. If it were not for their inborn laziness, I am sure the natives could find plenty of places abounding in gold. But they are too scornful of wealth, or too lazy—most likely the latter. I have seen a native walk through a fire because he was too lazy to go around. And they never bend down to pick up anything —they pick up the object from the ground with their toes and pass it to their hands.

Of the various suggestions, we eventually chose to move on to the MacIvor diggings. We heard that there was a rich surface soil called professionaly "serois". To get gold out of such soil special tools were necessary, so we bought a Long Tom and to transport it to the diggings purchased a cart and a fourteen-and-a-half-hand horse for £113.

The road to MacIvor is through the bush, just like any in Australia, and over the Campaspe River. The place itself is pleasant enough, the diggings in a wide gully and the tents all over the surrounding hills. There were only about 500 miners there and their earnings were very poor. We could not even earn our keep and with the end of 1853 decided to move to the Jim Crow fields where, according to the news from some of our countrymen, yields were satisfactory.

The previous year had been very difficult. We wondered what the new one would bring. Would I be able to live in a house again, because until then the sky, the meagre shadow of a tree, or a leaky tent had been my home. This hard life made each year seem overlong, particularly when the unknown future indicated a possibility that my stay in the bush might last for a long, long time. When I thought with longing about snow and frost in my homeland and realized that I might be laughed at for such sentiments, I had a ready answer: "Go to Australia, spend a year in the mines and then judge."

We walked toward MacIvor along the Bendigo road and arrived at Bendigo on the second day. We considered trying our luck there, but the hot season dried up the water so much that there was none for washing the stuff and even drinking water was scarce. We stopped for two days to attend to some business and I talked to Hodget who, after parting company with his Polish mates, had opened a shop in Bendigo with his brother.

Merchants are always well-informed about new diggings and conditions because it is in their interest to move to places where there is no competition and trading is good. They glean information from their varying customers. Hodget knew about two favourable spots, Mount Malaga and Jim Crow. We knew that some Polish miners worked at Jim Crow and therefore tended to favour it. On the other hand, according to reports Mount Malaga seemed to be a richer field.

We left Bendigo in the morning and pulled up at noon for a rest. Before we were ready to move off again Hodget and a young American doctor came along. Both of them, in their shirt sleeves, had decided to surprise us and have a cup of tea with us. We were a jolly company and time went quickly. When we were eventually due to move off and say goodbye, one of my companions said in a joke:

"Come with us to Jim Crow."

"All right", replied our guests without hesitation, and joined us. Only after we had covered a few miles did they realize they had said nothing in Bendigo about their intentions and that they had nothing warm for the night. We soon fitted them out with warm woollen shirts and blankets, but for a few days the search for our eccentric friends went on in Bendigo.

Late on the third day we reached Jim Crow. We found Stan C. with a few other compatriots of ours and recognized immediately that things were quite bright. They looked prosperous and happy and each had already put some gold dust aside.

We were too large a company to work together, because the shafts at Jim Crow were shallow. We sold the horse and cart to Anton B., who decided to earn his living as a carrier and the rest of us split into teams of two or three. We put our tents on top of the hill between two deep gullies. Around were tents belonging to miners of various nationalities: Dalmatians, Italians, French—mainly from Mauritius once called "Isle de France", and from the island Bourbon; further on Swiss, Spanish and German miners. In the evenings we listened to the songs of the musical Italians, light songs of the French and the occasional gutteral singing of the Swiss.

I preferred the lay-out of this mine to all the others I had seen so far, possibly because of the wilder nature of the hills and surroundings and possibly because the miners were mostly Europeans, to the exclusion of the English, and there was a better spirit of comradeship and co-operation.

The name of the goldfield Jim Crow stems from the name of a man who lived in the vicinity about fifty years ago.[3] He was probably a sailor who deserted his ship, or a convict who ran away to avoid punishment. In the early days of settlement the punishment for insubordination was hanging. Two prospective victims didn't wait for the execution but swam to the shore and disappeared in the bush. Very many years later an old man was noticed amongst a group of natives in Sydney.[4] He stood out for his different colouring and body shape. He ate the same food as they did, walked about practically naked, and at first could not speak English. However in a short time his power of speech returned as did his memory. He mentioned the name of the ship on which he had arrived and it was calculated that he had spent about thirty years with the natives. When his story was finally untangled, it appeared that after stumbling through the bush unable to find food, the two sailors fainted with exhaustion. One of them died, but the other was cared for by the natives and accepted into the tribe, and remained with them. At first he was afraid to remain in Sydney because he thought that the old verdict was still hanging over him, but when he was handed an official paper to the effect that the verdict was annulled, he decided to leave his native friends and remain in town.

Contrary to this story, there is no legend about Jim Crow except that he lived and hunted in the district, not realizing that he was walking over a fortune in gold. The deposits at Jim Crow diggings were not rich because the gold was widely spread in small quantities underground and no great riches could be gleaned from any one shaft. But nearly every miner managed to find a share to pay his expenses and save a little. Life was much more pleasant and steady at Jim Crow, where everyone knew that he had no chance to get rich quickly but that his work would bring returns. I began to dig at Doctor's Gully, so-called after a Swiss doctor who first dug in this vicinity. There was some gold in the two shafts I sank, but it was too far from water to make it a paying proposition, so we moved to the other side of the gully near a tree called a cherry tree, with leaves similar to cyprus, dark red berries and extruding a bloodred gum when the trunk was cut with an axe. Most of the native Australian trees are very different from ours, for they don't shed their leaves in winter. However in autumn gum trees shed their outer bark which hangs like ribbons, makes a peculiar noise in the wind, and being dry is suitable for kindling.

It took us only two days to bottom the new shaft for it was only ten feet deep. The test was good and we began to work the shaft, although with some considerable trouble because water was seeping in from the bottom. Every few hours we had to bail, but even so I always left the shaft at the end of the shift wet through. It left me with a neuralgic toothache which returned every spring for a few years. We worked the shaft for several weeks, according to the following routine: for two days I dug and sent the washing stuff up top and on the third day I washed. One day I was busy hacking away in the shaft when my companion called down that evening was approaching, that the others had left the workings and it was time to go home. I had only a little bit more to clear and wanted to finish it so that I could start washing next morning, so I decided to stay until I brought the last few bucketfuls to the surface. The bottom was uneven and one had to follow those contours of the bottom's surface on which gold is usually deposited. To gather the last bucketful more conveniently I turned in the shaft and moved my candle accordingly, when suddenly a glimmer caught my eye.

As the old hand which I had become, I had given up all dreams of getting rich quick and of finding a bonanza, but I bent down to examine the glitter. A small heap of grains glittered on the ground. I bent down and gathered the grains into a kerchief, my hopes rising that the following day I would find more little heaps. I didn't wait to collect the rest of the washing stuff, but went with my companion to wash out the contents of the kerchief. We got three and a quarter ounces of gold.

Unfortunately the shaft did not live up to my expectations and did not show any more heaps of gold, although we managed to wash out our share in the normal, hard, tedious way of the miner.

About this time our countryman S. got sick. This is a real misfortune in Australia, because there is no one to tend a sick person. Each hour taken away from work is costly. There is no scarcity of doctors, but they rarely occupy themselves professionally on the goldfields. Most of them are miners and I suspect that very few are real medicos, for there are too many claiming to be members of this profession to make it possible. Probably a few of them are good doctors, but how can one tell which one to select? They all charge £1 for pulling a tooth and the same for calling at their tent for advice which they call a professional visit. In any case most of them are interested in mining, not in curing people. So it is easier to be sick for a short time and die quickly than linger on and pay money to these men. Also there are no apothecaries on the diggings. There is no control over this profession and anybody can become an apothecary.

We looked after our friend ourselves and shifted his tent from the shaded, damp side to the top of the hill, where he soon got better and left for Melbourne. There he obtained a suitable position, being a qualified engineer. I believe one of the reasons for sickness on the goldfields is bad diet and lack of vegetables, or at least the very inflated price of vegetables.

When my shaft stopped yielding a good return, I went a few miles further to a little stream where only Poles and Frenchmen were digging, for all good places were taken up. Only a spot on the side which I claimed was available. My neighbours laughed at my choice, but I began to dig. It goes to show how little miners know about selecting their claims, for my shaft was the best along the stream. While there I made friends with Mr Barthelemy, a Frenchman, a sea captain and son of General Barthelemy. Captain Barthelemy was once the owner of a brig called *l'Epervier*, but some sad personal experiences caused him to decide to come to Australia.

By then the rains were continuous and flooded all the shafts. I managed to salvage our tools out of the shaft, but before I could go down again to lift out some bags with bottoming earth, all the shafts in the vicinity collapsed.

After the hold-up of the gold escort some time before, and the capture of many of the bushrangers engaged in this escapade, things became quieter and safer. But later we received news of many new hold-ups. The government had insufficient forces to control lawlessness in the bush and the situation in the city was not much better. I heard recently about the sad fate of Bishop Medianski. After the Hungarian uprising in 1849 the good bishop went to France and then came to Australia where he worked for a time on the diggings. He had to go to Melbourne to attend to business for his mining associates. One evening, passing only a few hundred paces from the barracks of the mounted police, he was set upon by a bushranger and shot at. Although wounded, the bishop knocked out the robber and went to seek help. His left arm had to be amputated as the result of this incident. This is just one example of the situation in the city, where frequent murders took place.

9

Melbourne, colonial politics and Eureka

By this time I had spent about seven months at Jim Crow and the newspapers began to bring exciting news from Europe. We waited impatiently every Sunday for the postman's horse to appear. As soon as we heard the hoarse voice calling "Latest English News" we used to run out of the tents and pay 2/6 for a slim newspaper, because any news at all would break the monotonous life of the miner, especially news from Europe and the theatre of war.

The newspapers were so popular that the sellers sold them for a tenfold price increase. The fate of Sebastopol was followed in the papers and continuously discussed. We discussed the general situation, the strategy and the mistakes of the various generals, and went over our own battles. In one of the papers was a short mention of amnesty for Polish exiles. There were eight of us on the Jim Crow fields and we all decided to return to Europe. We advised our compatriots on other fields about our decision and as soon as we exploited the shafts already begun we sold our tents and tools for half price and started to walk towards Melbourne.

How different the countryside looked to us from when we first walked towards the gold mines. We looked upon it now with a different eye—the grass and the bush seemed so much pleasanter, the buildings less primitive than when as new arrivals we walked for the first time through the bush into the unknown. On the third day we left the dense bush and could see the smoke from the chimney stacks of the steamers in Port Phillip Bay on their way to Geelong. Perhaps the steamer on which we were to embark on our journey home had already dropped anchor in the bay. As we came nearer to Melbourne the environment became more civilized—wheat fields, whitewashed houses, hotels. Progress moves fast in this city.

Our main aim was a decent rest, which anyone would deserve after two years' hard work in the mines. We looked for a boarding house and six of our company found accommodation with Mr Mercier, the Swiss whom we knew in Forest Creek and who had received some money from his father and opened a hotel in Melbourne. One compatriot and I could not be accommodated there so we had to seek lodging elsewhere. We were warned about a rat plague in Melbourne, but as a first class hotel cost £3 a day we went to Hotel du père Antoine, the owner was once a ship's cook. The place was reasonably clean, but the company in the bedroom was very mixed. In the large room in which we were given accommodation were ten beds. I had a bed on the side, next to me was my companion; then a Frenchman, M. Lombard, originally from Paris but recently from the Island of Bourbon; next in line was Mr Lopez, son of a general, who during some insurrection in Spain was a colonel in forces opposing his father's forces and thought it wise to put an ocean between them at the end of an apparently unsuccessful escapade; next was a young Jew—a commission agent who did not worry about the ritual restrictions but ate everything served at meal times; on the other side was a Frenchman and his mate who were engaged in the painting trade although they had never been painters before. The last lodger in the room was a cook who had a reputation as one of the best cooks in Melbourne and originally earned £7 a week, but drink got him and he now seldom worked. Antoine kept him free of charge out of sentiment for a fellow cook and countryman.

We waiting fruitlessly for a whole week to get some substantial news about the European situation. The ships arriving in port had nothing of concrete interest for us and it would have been folly to risk returning to Europe where worse hardships than hard work—a complete lack of it —could be our fate. Life in the Australian bush was much preferred by people like us to the hopeless vegetation in Paris or London, and so it was back to the bush for us.

My friends left Melbourne immediately, while I

remained another few days to try to arrange another method of making my fortune. I formed a partnership with my countryman M.S., to open a shop on the diggings. During those few days we purchased the necessary goods. M.S. owned a cart and we hired two more and loaded all three with goods purchased for £200.

Back we went, with some regrets on my part, not for the pleasures of the big city, but because this return to the bush could be for a long time. But life being what it was, I only gave a sigh and began to walk behind the carts towards Maryborough.

The distance from Melbourne to Maryborough is 120 miles and we arrived there safely without our horses being stolen or our carts breaking down. Our only mishap was to get bogged in a stream. We had to carry bags and boxes on our backs to lighten the cart before getting it out. Stephen S., another of our countrymen whom we engaged to help us and who went ahead of us to select a good place for the shop, had our tent ready in a good spot amongst many miners' tents. He was smart about selecting the place and erecting the tent before other merchants got to it. It is important to have a good spot for a shop, where many tents are erected, to get the benefit of good profit in what may be only a season of several weeks until all the shafts are cleared out, or too many competitors set up shop close by.

New shopkeepers put up a long flag on a pole as a sign that a shop is operating, and to gain custom treat the men to whiskey and wine, and women to confectionery, although many women like a glassful as well. This custom is widespread and well-known, so as soon as the flag is up crowds gather to buy their requirements and to get free drinks. If a new shop opens in competition to one already established, the already established shopkeeper is obliged to lower prices to meet competition and to offer free drinks as well, or he would not do any business.

There are two ways of buying goods in Melbourne. The good quality wares come from the ship stores at the Melbourne wharves at a slightly higher price. The cheaper way is to buy goods somewhat damaged by sea water, spoiled by lying too long in stores, or goods hard to dispose of, at auction. Usually the miner, in many cases an ex-sailor or an ex-convict, is not choosy and prefers goods which are cheaper.

There are also two classes of dishonest merchants. One type buys goods to the value of say £50 in the warehouses a few times running until he establishes credit, then takes a few hundred pounds worth of goods on credit, sells it cheaply and quickly on the diggings and disappears. It is difficult to find such a specimen in the wild Australian bush. The other type keeps books so manipulated as to show losses and officially and quite legally becomes bankrupt. This is not considered a shame in England and the colonies. On the contrary, a bankrupt has less difficulty in obtaining credit, because everybody expects that he has a lot of money put away, whilst an honest merchant may be embarrassed for ready cash to meet his obligations on time.

Another problem facing a merchant on the goldfields is a very rich competitor who is prepared to establish a shop in the vicinity and undersell until his poorer competitor is forced to move elsewhere, leaving the field to the wealthier man. An additional difficulty in trading is the dishonest customer. When a miner is digging a shaft in a promising place and runs out of money, it is customary to give him credit for food and tools and sometimes carry him for a few weeks. It happens only too often that the miner finds gold and disappears without paying. In our shop we carried all the basic necessities of a miner's life: coffee, tea, sugar, flour, all sorts of jams and tinned fruit. Tinned meats, fish and vegetables were carried in great variety as well as sausage, hams from Westfalia and bear hams from America. Sometimes we had cheeses from Switzerland, Holland and England, suits for men, shoes for ladies, lollies, needles, thread, cotton and silk piece goods, tapes and corsets, ropes and mining implements, nails in all sizes. We sold some basic drugs, including those which were useful in curing the after-effects of alcoholic indulgence. Naturally we had wines, spirits and tobacco and cigars. We used to buy bread from the bakers, getting a dozen loaves for every ten loaves paid for. We bought barrels of water and sold a quart for three pence, making ten shillings on a barrel.

I admit that I wasn't a good merchant by nature, for I didn't know how to price my goods to sell the more expensive lines; yet I was in charge of the shop. My partner was continually going to Melbourne to buy supplies. A round trip usually lasted ten to fourteen days. Stephen was best off, being on wages and getting £3 weekly and keep. He didn't run any risks. For the privilege of running the shop we paid a licence fee of £15 per quarter and receipts were made out to: Korr. & Co. storekeepers.

Our shop prospered, for we had a turnover of £6 to £12 daily and our cost of living was minimal, our only expense being meat. Profit from the sale of bread was usually sufficient to pay Stephen's wage. If it was not for losses due to giving credit and to the expense of treating the ladies to wines and spirits in order to get their goodwill, our profits would have been very good. As it was there was always some profit, if not great, because a glass of good wine or spirit sold for a shilling each and some of the ladies after treating insisted on buying a drink for both of us, which implied standing them another free drink. The ladies used to down this third glass with the words "my respects to you" and with such energy that all the flowers on their hats shook, after which they purchased many things they would not buy otherwise.

There was another side to dealing with ladies, I was the receiver of all the gossip on the goldfields but at the same time some worthwhile information came my way about the richness of individual shafts and the movements of miners away from the field.

After two months of this life, which did not suit me personally, most of the shafts gave up their gold, many miners and their ladies left, and that was the end to the profits. We had to look for a better place for our shop.

We found a gully near a stream which seemed likely and in three days shifted our shop there. The new place was not a patch on Maryborough, except for the number of thirsty ladies, because there was another shop nearby on the other side of the stream. About a month later the owner of this shop was held up by bushrangers, shot in the arm, and his cash taken. After that I began to lodge all our cash with the commissioner, because although I had a very sharp bitch and a handy revolver the risk was great.

By then I had had a three months' rest from hard work, but I was not a shopkeeper by inclination so I decided to give it up. After settling accounts I was happy to discover that I'd made a clean profit of over £25 in those three months.

I then teamed up with a Frenchman called Robert le Diable, but I soon found out that he was very lazy and in this lived up to the reputation of all French miners in Australia. We took a countryman of mine, B., into the company because he was a good worker, and we sank a shaft in the White Mountains. We went as deep as sixty-four feet without much help from Robert, and didn't find a trace of gold. So we parted company and I went with B. to the Alma fields.

The Australian Alma was luckier for us than the Crimean was for the British. On New Year's Eve I was up on top and my companion, who was digging, called out that he had bottomed. As a joke I called out "That's not good enough, show me some gold. We need some for the New Year."

"Here you are", replied my companion, handing over a nugget worth about £1. Solid lumps of gold are usually an indication of a good shaft, so we went happily to our tent to celebrate the New Year.

We began the year 1855 in a happy mood and in the company of a few countrymen of ours who came to the diggings and pitched their tents next door. The day passed in conversation and recalling our families and friends left on the other side of the globe.

Our shaft was very good and the first cartful of soil washed out brought us gold to the value of £18. Our neighbours however were unlucky, for every shaft they tried was a failure. Eventually we took them into partnership with us on our shaft. Our company thus rose to five members, but actually only four worked for a good part of the time, because I injured my arm. Later, when I recovered, I worked the winch, a job which with difficulty I could do with one hand while three others worked down the shaft and the last one shifted the earth brought up into heaps.

We had good returns from the shaft, but the work was exhausting in the heat and water was very scarce. Miners have various methods of quenching their thirst—some mix effervescent powders with water and sugar, others drink lemonade, some prefer beer and again some claim that spirits quench their thirst best. And small wonder, for the tavern tent is large and cool with multicoloured bottles on the counter, crystal clear glasses, comfortable benches under the wall in the breeze and a trim, neatly dressed hostess. The drinks are only sixpence each. The hostess greeted us pleasantly and told us that her mother was English and only English was spoken at home, but her father was Polish. Asked about his name, she said "Isaacs". We understood that her father, like so many Polish Jews migrating to England, adopted a surname made from his Christian name and an addition of a letter 's' on the end. And so Isaac became Mr Isaacs, Abraham became Mr Abrahams, Shloma became Samuels.

We finished work on our first claim and marked out a new one. This one was just as good as the first but the increasing heat dried up the water holes so much that there was barely water to drink and none to wash the dirt. We carried our earth to a stream a few miles distant to wash it. The stream was nearly dry, but there was water in some deep waterholes. We went there in rotation two at a time, while the other three remained digging. On one such occasion I killed a seven-foot black snake which was about to attack our horse.

I noticed that in summertime the natives seem to migrate into Victoria but as soon as the winter rain season begins they disappear. They probably migrate to warmer parts to the north. Although there are no frosts nor snow in Victoria, the damp winter climate could be too cold for the natives without clothing. About this time numerous groups crossed the bush in the vicinity and some camped for a day or two while others passed through, slowly followed by their numerous dogs. The natives keep them as a reserve so that if there is no other food they eat them.

It is not very pleasant to have a blackfellows' camp in close vicinity. On a moonlit night there is no rest for they make a lot of noise and scream in a peculiar way. Possibly it has a religious significance—a worship of the moon perhaps, or it could be that the *crodje* was giving talismans away. It's hard to say. Tired as I was after a hard day's digging, I was not prepared to flit through the bush at night in the hope of learning some of the natives' habits and ceremonies.

I recall another incident from this rather steady period on the field. One night our dogs began to bark. As they did not bark at the trees, it was not possums, so we came out of the tent, lit gum tree branches for torches and found a large snake, which we killed. Possibly I owe my life to the vigilance of our dogs.

Dogs are a great treasure in Australia. They saved the life of many a miner and prevented many mishaps, due to their warnings. The average price for a pup was £1, more for a grown dog, and good dogs fetched £2.10, £5, and even £10. I got my "Vistula" from an old miner when I was running the shop. She was only a little pup, and I paid for her by giving the miner a bottle of brandy and a bottle of gin. Vistula was a good dog and a good friend, but I had to leave her with a countryman when I finally quit the diggings. It was a difficult parting.

Those were days when the whole of Australia expected many changes and improvements in the administration of each of the colonies.

Governor La Trobe, richly rewarded, left for Switzerland, and some time later we saw news in Melbourne papers that his wife died as the result of a sickness acquired during the sea voyage, and that the governor married again.[1] A new administrator left England to take over the colony.[2] News about his justness, energy and ability travelled ahead of him. The events proved that one may be an excellent admiral without the gift of good management, particularly in a new colony where so many things have to be taken into account and where it is not sufficient to be either very mild or very severe.

Laws in the colonies were different from those in Great Britain because they were especially designed for the various possessions of the Crown. Some of the wise and gifted inhabitants of Australia, and there are quite a few, realized the importance of this land and its specific and different conditions. They felt that a land 15,000 miles away from the centre of the Empire has to be governed in a way suitable to the local needs and demands a constitution suitable to those needs.

Those people air their views in newspapers and periodicals and are supported by a great number of settlers and an even greater number of miners. Numerically they form a large section of the community, too great to take patiently the will and whim of one man used to command yet unaware of the conditions needed to bring prosperity to the colony, and therefore to England. A man is a bad choice for such a position if he is unable to accommodate to circumstances, because the expression "it cannot be done" does not exist in the military lexicon and consequently in his. In the early days when all the white population of Australia consisted of a few hundred convicts and the military garrison was large enough to guarantee that the will of the governor was paramount, it was probably necessary to mainly consider the element he had to directly supervise. Now the majority are decent, thinking people from different countries and walks of life. They have their own point of view and realize that they can insist on it, because the government has not the forces to press them to change their views. They are the type who cannot be coerced and will not obey a command. They are like wild, fast horses freely roaming the expanse of the Australian bush: apply a hard hand to them and they are liable to go the way America went and possibly soon. We shall shortly know if the new governor realizes the true situation.

Preparations for the welcome to the new governor in Melbourne were elaborate. Decorated arches, illuminated at night and carrying words of praise and welcome abounded. Statues of Minerva and Bellona, flowers and addresses were readied and waiting for the day. But the ocean has slight respect for governors and the steamer carrying His Excellency Admiral Sir Charles Hotham, Governor of the Colony of Victoria, arrived two weeks late.

A railway linking the port and the city two miles distant was used by the governor and his lady to arrive in Melbourne. Here they were practically bodily carried to the carriage by the enthusiastic populace, and after the horses were unharnessed they were pulled by manpower to their destination. Speeches, songs, addresses, balls and receptions lasted a whole week.[3]

A waiting period took place, during which the new governor undertook the task of getting acquainted with his domain. He visited towns, both those already in existence as well as towns in name only. He visited the mines and inspected them thoroughly. He inspected the horses and stables of the constabulary. He inspected the accounts for the upkeep of the horses and the stablemen. He received petitions from miners, settlers and townspeople. Eventually he ordered a brick fence erected all around his summer residence some three miles from Melbourne in a beautiful location near the sea, and caused a large ballroom to be erected there as an addition to the building.

After a few months of meditating, His Excellency reduced the government clerks' pay, took the free servants away from them, and reduced the allowance for horses to the commissioners of districts, forcing them to cover their districts on foot. He also reduced the number of clerks and suspended others for even the smallest misdemeanours. Instead of enlarging the police force, which was necessary to ensure public safety, he reduced its strength. He did not reply to requests and petitions, and would not grant the miners representation on the Legislative Council (advisory). He left the councillors in no doubt that he would govern without their help. Naturally this increased dissatisfaction all around.

Most of the public works to improve the roads ceased and forests were parcelled up and sold to the highest bidders. Eventually the newspapers and the public

Golden Point, Ballarat. 1851. This engraving by Thomas Ham was made from a drawing by David Tulloch and was originally published in *Five Views of the Gold Fields of Mt. Alexander and Ballarat in the Colony of Victoria*, 1852.

completely changed their attitude towards Sir Charles, from whom so much good was expected. This dislike kept growing, meetings of protest were called, and everywhere preparations were made to oppose an unpopular governor.

The goldfields were full of agents of this opposition recruiting new members.[4] An uprising was projected to begin in a number of places simultaneously, but either the date had not been announced, or the densely populated Ballarat fields could not hold the feelings in harness any longer and the action against the government began there.

But as always, where there is a lack of experience, what should have been action became a child's play at soldiers. About 200 miners selected a hill to fortify, erected a barricade, established sentries and provoked passing patrols of mounted police until the commissioner received military reinforcements from Melbourne.

Although the miners knew about the reinforcements they did not change their strategy nor the locality of their encampment. Their sentries continued to be posted but the rest occupied themselves in eating, drinking and singing revolutionary songs, after which they went to sleep. The commanding officer of the government forces picked the time for attack to be well after midnight. The miners, secure in their belief that they would not be attacked, did not at first believe the sentries when they were woken with the news of an attack. The bullets whistling past forced them to reach for their firearms, but half asleep and panicky, many of them shot wildly and had no ammunition handy to reload, because in their haste they had left it in the tents.

The controlled salvos of the attackers did much more damage than the individual shots of the miners. Several miners were killed in the first salvo and the others lost the will to fight and ran away. But the experienced commander had another detachment ready at the rear of the camp and these now opened fire. Many miners surrendered, to be tied up and placed in a special encampment. In the general shooting a few women were shot when bullets penetrated their tents.

Altogether over twenty were killed and buried the following day. The imprisoned miners were driven to

Melbourne for a trial which was followed with great public interest and suspense. A few weeks later the papers printed a list of those condemned to die, and in the same issue printed protests from towns and mines against the verdict. In Melbourne a few thousand people attended a meeting and petitioned the governor. Apparently the submissions were sufficiently good to influence the governor's mind and the accused were pardoned.

Soon after, by an edict from England, Sir Charles was elevated to the position of governor-general of the Australian Colonies.[5] But these occurrences did not make the Australian people well disposed towards Hotham. No reforms improved the situation in the colony and the blooding in Ballarat did not subdue the impatient. Everyone awaited news from England, where a plan for a new State constitution was submitted, one more suitable for Australian conditions. It was rumoured that the Queen had approved a new constitution and sent her approval to the governor.

The Legislative Council demanded the proclamation of the new constitution, but for reasons never given, His Excellency the governor did not think it necessary to do so. One after another the ministers refused to serve. Sir Charles was adamant and took no notice of this nor of a wave of sharp criticism in the daily press. He formed a new ministry but actually governed by himself and according to his views only. This state of affairs was suffenly altered by his death due, according to the press, to inflammation of the brain. After the death of Hotham, the old general MacArthur temporarily took office.[6]

There was much talk that under the new constitution Australians would vote for governor and would probably select one of their own for this position. But I have never read the constitution and don't know it. The press commented widely on it and there were continuous discussions on the subject, but right up to my departure MacArthur continued to act for the Queen. Something must have changed, however, because the citizens of Melbourne formed a national guard, which trained once a week. For what purpose I have no idea.

10

Shopkeeping, the traffic in women, and quartz mining

I worked at Maryborough and later at Alma when all these occurrences took place. We had two good claims there, but when those were finished we sank a number of shafts unsuccessfully. Sick and tired of it, we went to a spot still on the gold line, but where no shafts were in the vicinity. We drove a shaft to a depth of fifty feet when we struck a greasy, whitish earth very much like pipeclay, which is usually at the bottom. It is a very confusing earth, because when it dries up it looks like pipeclay. We went on digging and came across a layer of gravel. When other miners noticed we came across gravel, they marked claims nearby. By then we were sixty feet down.

At that point two of my compatriots, W. and B., bought a third of a claim in a mine in Quartz Mountain, three miles away from Alma. This was the time when gold began to be successfully extracted from quartz. W. proposed that I join in this enterprise and also partner him in a shop which he intended to open on the Quartz Mountain field.

I decided to inspect the prospect first and calculate possible gains. It seemed suspicious to me that the English group working the reputably rich claim would be willing to sell a third of it for £300 to Poles from far-away diggings when they had English and German neighbours. The story my friend told me was that Mr Foster, who came from Jamaica via a stay in England, became a postmaster on arrival here. He had thrown in his job to buy the claim with two partners. One of the partners was not suited to the hard work on a quartz field and wanted to sell out. Mr Foster had approached my friend a week ago to buy this share.

We walked to Quartz Mountain. Here huts and tents neatly set out gave signs of the prosperity of the miners and of their intention to remain on the spot for a long time, possibly several years. The fact that they did not seem to work very hard—I saw many walking about smoking their pipes—I took as a good omen. The miners seemed to feel assured of good earnings, so a shop in a place like that should give good returns.

We next went to see the claim. Mr Foster waited outside the shaft. Dressed as any miner, he had at his belt a bag for matches, a tobacco bag, a pipe and the usual miner's knife in a leather sheath. His talk was full of praise for his claim and he invited us to go down and inspect it for outselves. The shaft was second in the row, for the first belonged to a Mr Cooper who employed a Negro worker. Foster's was next, much deeper than Cooper's, and the third belonged to Mr Nightly, an ex ship captain. From the depth of this claim I could hear the barking of a dog. Mr Foster explained that Mr Nightly was resting that day, but because quartz is valuable, he always left his dog to guard the place in his absence. Later I learned that this was a clever trick.

I had to use a rope to get down the Foster shaft. There his partner Atmuir showed us some broken-off lumps of quartz with gold specks in it and claimed that the average should be 9 ounces per ton or better. We were invited for a meal at Foster's tent, where we met his wife Grace. Neither the tent, the poorest and smallest on the field, nor Grace's manners and liking for wine made a good impression on me.

I was reflecting on all these points on our way back home when W. happily remarked on how lucky we were to be offered this opportunity.

"You are the lucky one", I retorted, "because I will not buy into the partnership. I don't believe that your English friends would offer you a bargain if there was one there. As far as the shop is concerned it seems a good proposition and although I won't go into partnership with you, because I don't know how long I intend staying in Australia, I'll lend you money to start one and I'll work in it for you on £3 per week if you repay me a portion of the loan money weekly to clear the debt in six months time.

In the course of the conversation it came out that B. had no money and I was put in a situation where I had to agree to lend W. another £100 to buy the share in the mine. Also I found towards the end of our walk back home that any profit from the mine could be expected only after a steam engine for crushing the quartz was installed on the fields.

I left my half-finished claim at Alma to my partners and became a shopkeeper again. They found gold, but the claim was not as good as some of the neighbouring claims. Thousands of claims were dug in the vicinity with good results. Because we were the first to select the locality the place was named Polish Flats. In years to come perhaps a Pole travelling about Australia will find a trace of the earlier presence of his countrymen in this spot.

One week later we opened a very decent and well-provisioned shop at Quartz Mountain. As usual everyone called in for free drinks and a look around. The most frequent visitor was Grace, who was always running in to give a hand. Our shop was divided into three parts. In the first part was a counter with scales, shelves with all sort of goods and bags of sugar and chests of tea, etc. piled at the side. The second part had the walls covered with a woollen material with a floral design and two tables with benches in the middle for entertaining and treating guests and valued customers. The third room was my bedroom and a storeroom for wines, spirits and vinegar. In this room on an upturned barrel stood an array of bottles of wine and spirits earmarked to the expense account for treating customers.

Grace got into the habit of sitting quietly on the bench in the middle room, but every time there was a lot of customers in the shop and I was busy, she used to run into the bedroom. I noticed that after each such visit the level in the bottles was down. Eventually I caught her drinking a glassful of brandy. It was then that I asked her not to help me again and not to go behind the counter again as I found her visits too expensive.

Some time later a young, decent-looking miner walked into the shop with W.'s promissory note given to Foster for the share in the mine on condition that he would hold it until the due date in his possession. It appeared on interrogation that the young man sold Foster and Atmuir half of the claim for £200, not one third for £300. Consequently W. felt very put out by the situation, particularly because Foster and Grace had already run up a large account in the shop which we felt would cover part of the promissory note. Caught in a bad situation, W. and B. pestered me again to become a partner in the mine, which after reminding them about my original opinion about this proposition I reluctantly agreed to do in order to help them out. On one condition, however: that the last promissory note still due in a month's time would be taken away from Foster and given to Atmuir, who in the presence of witnesses was to agree not to pass this note to anyone.

From the report of the young miner when he came to collect his money on the first promissory note, it appeared that we had been badly taken in. He told us that he sold his share because the gold was petering out. A month later W. settled accounts with Atmuir and got the last promissory note from him. Foster, when he found out about it, came the next evening to the tent and tried to bluff the promissory note out of me. But I placed a life preserver on the counter in clear sight of Foster and told him to leave. He left, but a few days later I received a summons to appear in court in Maryborough.

I had to walk the four miles to Maryborough in spite of a festering old wound of mine in the leg. Because of my condition I was given a seat in the jury box by the chief magistrate and from there listened to Foster's testimony. When my turn came, I apologized to the judges for my faulty English and testified as follows:

1) The promissory note was lodged in my safe keeping by W. and not by Foster, just as a number of watches and other valuables were lodged with me by various miners. My understanding is that I could not give the note to Foster because he did not leave it with me.
2) Foster claimed that I was rude, but he was on my premises and was rude to me. To his disrespectful words I replied only "Get out or I'll throw you out."

3) I had a life preserver and I admit that I would have used it in my self-defence if attacked by Foster. He came late at night, the time when only bushrangers go about their business, and I like to be prepared against such visits, hence the handiness of the life preserver.

According to usage, the jurymen were supposed to go aside to consider their verdict, but one of them said that he felt there was no need to waste time on that and the jury brought in the verdict of "not guilty" without leaving the box.

Apart from winning the case, the fortunate upshot of this affair was that I had now an enemy in Foster and was no longer bothered by his visits. Grace came to the shop once again on a Saturday night and began an argument with the young miner who was their original partner. She grabbed two jars of preserved fruit and threw them at him. I had to bodily carry her out of the shop and forbid her to show herself inside again. From that time I saw the Fosters from a distance arguing with each other, probably because they could not get credit anywhere for drink. Eventually one day when Foster returned from a journey lasting a few days he did not find Grace in his tent. She had left with another miner for some unknown destination.

This was not the first incident of this sort on the fields, nor should one wonder, considering the type of women brought to Australia. Originally deportation was a punishment for serious crimes, later vagrant and homeless women were deported,[1] finally a certain Mrs Chisholm carried on what amounted to a traffic in women.[2]

Several years ago Mrs Chisholm, wife of a ship's captain, arrived in Australia. As Mrs Meredith writes in her work,[3] Mrs Chisholm was busily engaged marrying off or placing in service, women freshly arrived from Europe. Mrs Meredith praised Mrs Chisholm, giving her credit for happily settling the fate of many a young girl. It is possible that after settling in Australia the lady developed a wide circle of contacts with settlers, knew what each needed, or simply took specifications and orders from them. She had an agent in Melbourne, who used to send

Store Drays Camped on road to Ballarat. This lithograph by S.T. Gill was originally published in *Sketches in Victoria*, 1860.

the women to her as soon as the boat arrived in port. The women, not knowing where to go nor how to fend for themselves, gladly went where a secure future was promised. So Mrs Chisholm sorted them all and sent them on into the bush.

However, many claim that every recipient of a woman through Mrs Chisholm, be it a wife or a servingwoman, had to pay a tribute. This seems likely, for while Mrs Meredith says the conduct of Mrs Chisholm at the beginning was exemplary and she had the blessing and help of the government, things changed later.

Two years before my departure from Australia (this would make it 1854), Mrs Chisholm went to England to recruit women. Being well-spoken and of an imposing appearance, she found it easy to convince a great number of gullible young women that the gold-bearing country was waiting for them with the promise of success. Oh how many parents blessed with a large family joyfully anticipated good fortune for their daughters in the antipodes, so they entrusted them to the care of the benefactress who arrived from so far away and opened the gates to a rosy future for these poor and plentiful maidens. Having lived in England for a considerable number of years, they had lost hope in the future.

Mrs Chisholm, painting only the bright side of Australia, carefully omitted its defects. She never mentioned that numerous transports of women had already gone to Australia before the one she was organizing. Consequently not all of the new emigrants would be able to get married nor get an occupation they were suited for. Therefore they would be obliged to wait in an "emigrant house", a large wooden structure built in Melbourne for emigrants arriving without any means or contacts, and that they would have to remain there until rescued by someone. Yes, rescued, because to remain there is on par with remaining in prison.

True, lodging and food were free, but it was so carefully calculated how many ounces of food were necessary to keep a human being alive that women considered it a great stroke of luck to be rescued from the government. They were prepared to grasp at every opportunity and agreed to go with a sailor, an ex-convict, anyone, even without having met the man previously, and knowing nothing about him. Nor was there any difficulty in obtaining a woman from the emigrant house. Whoever wished arrived, picked one to his liking and carried her away with him after reimbursing the institution a few pounds which the government had spent on clothing, because usually the women had arrived in Australia in tatters.

Such a method of securing a future was probably not a blow to a woman living from day to day, without any moral or religious principles. But how could it be accepted by decent girls just out into the wide world? How difficult and shocking was the situation of these poor beings! Far from those who could advise them what to do, left at the mercy of the world, hungry, they accepted service in which they were expected to do things for which they had no training, or if they could not obtain an occupation and poverty pressed, they disappeared into the crowd. Rarely did they make good again and only then when they happened to find themselves in the tent of a decent man, or at least one not totally bad.

Lucky the girl who—not having found in Australia a lifelong guardian, or an occupation suited to her abilities—could earn enough to pay her fare home by honest means and return to her parents as unspoiled as she left them. In some cases the parents were able to send her the fare. In any case she left Australia sharing with thousands of others the conviction that while things might have seemed bright from afar, reality is often harsh. Good luck, or good fortune depends on accident, which neither hard work nor trying can bring about, for it is solely in the hands of Providence.

Three such girls were returning to Europe on the same boat as myself. Miss Ann M. was the daughter of a captain, while the other two were daughters of a merchant from Cardiff. The merchant had five daughters at home and earmarked two of them for mothers of future members of the Australian parliament. Unfortunately all such positions were occupied, the war in the East stopped further immigration of men, and it was difficult therefore to settle down satisfactorily. Fortunately the young ladies had brought complete outfittings with them, being certain they would need them in their new life. They had also some money, so they leased a small house in Melbourne and waited for two years, earning some money by manual work. Finally, seeing that every boat brought many more women and only a few men, and that prospects looked very dim, they sold their chattels and returned to Europe.

Originally the ratio of men to women was six to one, but the ratio may well be reversed. Gossip has it that the government has a large problem housing and placing the superfluous women. After feeding them for a long time in the emigrant house in Melbourne, the government eventually sends them in large batches of several hundreds to the diggings, thus saving the miners the time and expense of going to the capital city in search of a wife.

There is talk that a few hundred women were sent to the Fiery Creek district and placed there in batches of a few per tent. In the township of Castlemaine everybody who wanted to take a woman and proved that he had £10 could get a woman and a dowry of a similar amount with a tent and two blankets. With a dowry like that one could begin digging. After all, many of us began digging without a penny. But who would choose to take such a bedowered wife? A drunkard, an ex-convict, or even worse—a bushranger? What a wonderful future for the poor girl if she is a decent person! Generally the other kind of woman were left to men who knew how to get rid of possible future troubles. After spending the dowry the men would abscond from the bare tent, leaving there the unfortunate women. Then these in turn would go to another district or colony, to marry again and repeat the procedure. Oh in Australia there is quite a crowd, both men and women, one no better than the other.

But what else can a government do with the every-rising flood of arrivals? By now nearly every miner has a woman in his tent, who has the privilege of calling him her husband. Over ten thousand Chinese live celibate lives but they have no inclination to turn their small eyes upon the

English beauties. In towns and settlements there are plenty of servants. Newly-arrived women on the whole are not suitable for agricultural tasks and are not used to heavy labour.

Perhaps if they were proficient in arms, one could form divisions of Amazons and send them to Canton or to bring order amongst the Sepoys in India. Alas, they are not suitable for either of those tasks. Truly I don't know what the government will do with them.

Mrs Chisholm knew the situation, but did not enlighten the girls in England to the fact that in Australia they would quite likely become temporary wives and would have to fend as well as they could. Instead she loaded hundreds of them on boats, charging, so I have been assured, £3 per girl. A very handsome income for Mrs Chisholm to collect, and the curse of many an unfortunate girl.

I expected the rule of Admiral Hotham to bring better order and ensure more safety to the life and property of the inhabitants. But none of this happened. Early in the morning I could hear shooting in the distance caused by claim-jumping by the so-called Tipperary Boys. It had been going on for several days—the rightful owners of the claims and the claim-jumpers potting each other as if it were duck shooting. On Fiery Creek there were hold-ups, attacks and atrocities. On Alma field only three miles distant from our shop a bushranging band under the leadership of Black Douglas made its headquarters. Officially they worked and dug on the field, but their main occupation was stealing from tents and robbing shops. The tents are left unattended in the daytime when everyone is out working, and the shops in the depth of the night when no one walks about.

The method was to organize a fight late at night in the vicinity of the front of the shop, and when the merchant came out curious to see the fisticuffs, accomplices would cut the canvas wall at the rear and grab what they could. The government could do little to stop the lawlessness because the police force had been reduced and the gold which could have been used to increase the force was sent to England.

The miners were forced to deal with the situation themselves by forming committees and armed detachments. They did so at Alma, where they discovered who the robbers were, surrounded their tents at night, tied them up and burnt their tents. Black Douglas, a huge solid man, resisted the longest and wounded one of the posse with a knife. Eventually himself wounded and overpowered, he was tied up, put in a cart and sent to Maryborough by an escort of a few hundred miners. Sent from there to Melbourne he was eventually hanged. So finished Black Douglas, deported for crimes to Australia where he did not give up his career as a murderer.

The robberies even began in Quartz Mountain. I saw a rider holding a revolver in his hand galloping past our tent. Although not certain, I had a suspicion that he was up to no good, but one cannot interfere with a man just on suspicion. I soon found out that the man approached two Chinese. One of them was holding their horse while the other washed some dirt. Pretending to be an official, the man demanded to see their licence. One of the Chinese brought out a leather bag in which, with the licence, he kept the gold. The robber grabbed the bag, knocked out the other Chinese who was holding the horse, jumped on the beast's back and galloped away.

About the same time two strangers arrived in the settlement. One, known as Red Jack, was red haired and always wore a red shirt. He was quite talkative and always laughing in a sarcastic way. His companion, who was dressed in black, was always silent and morose. The two erected a tent, had mining tools, but never marked a claim. When they disappeared for a day as often happened, they would return drunk. A number of horsemen used to visit them, some apparently just to pass a message, because they did not even dismount. After a while the two were joined by another companion by the name of Dick, who was an ex-convict. After their arrival a number of burglaries from tents and hold-ups on the road began.

I tried to discourage them from visiting the shop, but uselessly, for they professed a great love for Poles, would buy drinks and food in the shop and eventually, as always happens, ran up a small credit account. Although suspicions centred more and more on these three characters they were never caught committing a crime and nothing could be done. I was getting quite worried because they were in the habit of visiting the shop at all hours, when all the miners were at work. One day a young publican who had his tent hotel half a mile away walked into the shop and told me Red Jack had spent the night in his tent pretending to be drunk, and that in the morning some gold and a watch were missing. The publican was making the rounds of nearby shops to warn shopkeepers against acceping this gold and watch. He also told me that Red Jack, refused further credit for drink, cut the side of the tent and ran away with a bottle sneaked out through the slit. Just then Red Jack walked into the shop. When accused by the publican he laughed and denied the whole affair. As there had been no witnesses he felt secure that nothing could be done to him.

For a few days after that incident I saw no more of Red Jack. I discussed the matter with W. and we tried to devise tactics to use in case of an attack. But there was no attack. Business reasons made W.'s absence from the shop necessary and although he used to return home at night he was absent most of the day. Finally Red Jack picked just such an occasion for a visit. He came early in the morning asking for a bottle of brandy on credit which I refused to supply until he settled the outstanding debt. Just then a customer called and Red Jack quietly left the shop. An hour later, when all the miners were already at work, Red Jack came again, ostensibly to get some liquor but actually to check if anyone else was in the shop. Sent away again empty-handed he left quickly, which contrasted with his usual habit of walking slowly. I became anxious at this sign, expecting an attack, and started to look for something to arm myself. We had two shotguns for sale hanging on the tent wall, but they were useless, being rusted in the bore. I was just about to hang the second gun back on the wall when Red Jack entered with two companions and a grin on his face. I don't know if it was the sight of the firearm in my hand, or the fact that I immediately pick up the life preserver, or perhaps both which wiped the smile off his face. A short verbal exchange ended in all three sneaking out of the tent. Next day news arrived that a government commissioner was coming to Quartz Mountain and I saw Red Jack sneaking off with his belongings. His tent remained for a couple more days in its old place then disappeared.

The fact that the robbers departed did not calm my apprehensions. I was worried that knowing the lay-out of the shop and its rather solitary position they might choose a suitable time and visit me again in a more numerous company. Fortunately a few more tents were erected in the vicinity. This made things safer for me and the inhabitants made my last months in Australia really pleasant.

They were mainly young, educated and good-hearted people. Mr Butt used to be a government commissioner at the Sandy Creek field. Unfortunately during his time there a chest with gold lodged by the miners was stolen. One of the guilty men was caught and the gold recovered, but the governor sacked Mr Butt, choosing not to take into consideration the fact that his district covered thirteen miles and he was rarely at his headquarters because his total force consisted of only four foot and several mounted police, who had to patrol most of the time. Finally the fact that the gold was recovered and the culprit apprehended proved that Butt was innocent. But Mr Butt was forced to exchange his government commission for a miner's pick and so put his tent next to our shop.

Our next neighbour was young Mr Messenger, a happy-go-lucky well-built man, very well-educated and widely travelled. He had become too friendly with a lady at the Mount Couron diggings for her husband's liking, hence the change of location. The third tent was inhabited by two countrymen of ours—Anthony Bi. and Alexander Br.

I cooked breakfast and tea for all these because they were away all day working. On Saturdays we had the additional company of Stan C. and often Stephen, who as soon as he had accumulated sufficient capital started out on his own. Many others occasionally spent an evening with us and the conversation covered all subjects.

A very curious person who joined us occasionally was Kosmos. So he called himself and when asked about his nationality always replied that he was cosmopolitan. He knew all the European languages known to us and was a walking encyclopedia. To look at him no one would believe it—long dark hair covered by a hat full of holes, an old mining garb over which he always wore a blanket, only his dark eyes expressed his intelligence. I knew he was quite wealthy, because once he sent back £100 which someone sent to him. We were very friendly although I could never extract from him anything about his past or about his personal affairs.

Time went quite pleasantly in this environment and the day I was planning to leave Australia fast approached. But the matter of the claim shared with Foster and Atmuir complicated my departure. As it turned out finally, I lost £75 on the deal. I gave up my share in it and Foster manoeuvred my friends to lose their share in it by bringing a few false witnesses to say before the commissioners that they hadn't worked the claim for over twenty-four hours.

One of the witnesses was a competitor of ours who had recently opened a shop in the vicinity. This so incensed our young friend Messenger that he challenged the big bulky storekeeper to a boxing duel. To the surprise of all the watchers, the agile Messenger won. As the usual habit demanded, both adversaries came to the shop to have a few drinks together to show their sportsmanship.

After this I got very friendly with Messenger, who told me his life story, including his service at sea as a midshipman and then as a whale hunter off the Australian coast. This led to his landing in Adelaide where he met and became friendly with a young girl. This eventually led to a lot of trouble, for he overstayed his shore leave and as a punishment was kept for twenty-four hours sitting in the riggings without food and water. After that he attempted to swim ashore, but was caught and kept in irons for a fortnight until the ship arrived in Sydney. This treatment was too much for my young friend. He deserted in Sydney and for the last seven years had been working at all sorts of jobs until the discovery of gold.

The shafts at Quartz Mountain were dug deeper and deeper and some of them showed a lot of gold at considerable depth. One of the lucky ones was George Guthol, who eked out a very modest living with his wife over a long period. In between sinking his shaft he washed surface soil for a bit of gold. To help her husband his wife bought spirits in bulk from our shop and sold it illegally by the glassful, and took in washing from the miners. George, a decent fellow, sometimes asked for a glass of spirit on credit in secret from his wife, but never more than one.

It was surprising to see the change in Mrs Guthol when she came into money. No more the shy apologetic buyer, she tried to give the impression of a lady born to it, without really knowing how. Only the best was good enough for her, not that she knew what was good or bad, for the price of the article had to be the criterion. Even her way of speaking changed: she tried to speak out of the corner of her mouth, apparently under the impression that that's how high born ladies talk. She was a good customer and bought only in our shop, but always took a long time choosing and criticizing the goods. She also put a condition on us: no credit and no spirit for George, which somehow did not prevent George from satisfying his inclination to a drop of spirit. I estimated that his daily dose in those days rose to ten glasses.

Some of the quartz shafts reached a depth of 200 feet and water seepage became a problem. The water was salty, like seawater in taste, although the sea was about 160 miles away. It would take the miners all night and part of the day to bail the water out sufficiently to be able to put a few hours daily into actual digging. This was too much for the strength of the miners and consequently they formed larger companies to be able to relieve each other at this work. Even this proved insufficient, so there was talk of installing steam driven pumps. There was also talk of a permanent settlement and an increase in population to operate all this modern machinery.

Merchants and speculators investigated the feasibility of opening more shops and hotels. I took a rather pessimistic view of all these plans, partly because I was only weeks away from sailing back to Europe, but also because I remembered the fate of Foster's shaft which turned out to be a non-paying proposition. Couldn't the same apply to all or some of those deep shafts?

11

An end to exile

The last few months moved very slowly: I counted every day bringing me nearer to my departure. Finally the last week came and it was time to say goodbye. My friends and acquaintances knew about my plans and many of them came a long distance to farewell me and to try to talk me out of my intention to depart. My old friend Wilhelm Gayer came from the New Bendigo mine, where he was assembling a quartz crushing machine. He was at the time planning to leave mining and to settle as a farmer near Adelaide. He said to me: "Don't go to Europe, come with us to Adelaide. If you want to be a farmer, I'll give you a little house and a field there next to mine. It's good soil. You'll do all right there."

Next day Christian Schpötter came to say goodbye. He had parted company with Wilhelm a long time ago and still drank heavily. An old Englishman who had a farm in the bush and used to come every week for supplies on a small pony called me a fool for thinking of going away and proposed that I come and share his farm with him. But I told him that my ambition was to be eventually buried in my native land which he thought a great folly.

Finally Stephen came from Melbourne to see me and brought news about Kosmos. Three weeks previously he had stopped at Castlemaine, where a countryman of ours, St., was living right near the highway, engaged in making beer. He had as an assistant a Dalmatian chap, who used to be a ship's captain. It appeared that Kosmos' father, a very wealthy man, owned a fleet of ships plying the Adriatic. Kosmos had a very fine education, but apparently had a romantic affair in Paris followed by a disagreement with his father. At present he was busy equipping a ship to sail to the South Pole to look for some undiscovered lands. The Dalmatian was to become the captain of the expedition. The ship was to leave the shore in a week's time but the Dalmatian was doubtful about the success of the expedition because, as he put it, the British would not allow Kosmos to sail under the British flag and if he tried to plant some other national flag on a newly-discovered land British warships would use their cannon to blow him out of the water. When I got to Melbourne I could not find Kosmos anywhere to farewell him.

The last Sunday at Quartz Mountain was memorable. My friends the Scottish Highlanders came in their weird national garb, as did some English and Dalmatian acquaintances. George Guthol presented me with a piece of quartz showing gold specks as a memento. The next day, Monday, was the beginning of my way back to Europe —a return earned by long and very hard work.

Even on the Monday morning I couldn't believe it was the day I would leave the mines for good, leave this place of so many problems and worries, of so much back-breaking work, leave those ageless forests and bush with the natives wandering through it, part from my companions with whom I'd shared years of hard work and rough conditions.

But the coach was leaving Maryborough for Melbourne at 12.30 P.M. My friends did not go to work that morning, but walked with me to Maryborough to say their last farewells there. When the horses started at a gallop, I looked back once again but could not see any of them for the cloud of dust raised by the coach. The horses were changed every two hours, and we pulled up at Castlemaine (once known as Forest Creek) towards sunset.

The coach stopped there for the night and started again at five o'clock in the morning for Melbourne, after picking up passengers arriving from Bendigo and other fields. I used the time to visit the place, but could not find any traces of spots known to me. The growing town covered all those landmarks. Where the cemetery was previously, a wide street now stretches, flanked by houses with well kept gardens. I could not even find the spot where our own tent was pitched. In Australia people try to cover the traces of old history and natural landmarks, so that before one can blink an eye the scenery changes. Naturally the presence of gold speeds up the process.

The coach was long, roofless, well sprung, with benches across taking a total of twenty-four passengers with their gear stacked under the benches. We drove very fast, travelling eighty miles in one day, partly on made roads and partly over roots and pot holes, shaken about so much that I felt it for days after. Two miles before reaching Melbourne an axle caught fire and we had to pour water on it to stop it smoking. Only such an accident or the fall of a horse could stop the coach, for the driver was driving as fast as possible. This was the fastest pace I have ever experienced and I have had some experience in my life of driving fast.

We arrived in Melbourne in late afternoon. I ordered my luggage to be taken to a house with a sign "Polish Hotel", where I was to meet with S. who arrived in Melbourne a few days earlier and was to travel with me to Europe. The hotel was kept by a Polish Jew by the name of Silberberg. He had shed the traditional orthodox garb for the local style and until he spoke no one would have known he came from Poland.

His wife and elder daughter Anna spoke Polish, but the younger one knew only a few words because she had arrived in England as a baby. She became completely assimilated to local conditions as far as clothes and manner were concerned and had even changed her name from Esther to Estelle. Both sisters were good looking and well brought up. Their mother was deeply attached to the country of her birth. She often said to me: "One can live anywhere and find beautiful spots, but there is nothing anywhere to compare to our Warsaw. I pray every day to God that I may yet have the opportunity to spend my last days and be buried in Poland."

I was deeply impressed and had a lot of respect for this woman for her attachment to the country of her birth, to our country. Mr Silberberg was a good hearted man like many other Jews living in Melbourne and on the goldfields. When asked about his nationality he always replied: "We are Poles." Their straight dealings, their assistance given to anyone from Poland, their support for anyone in difficult circumstances and their general attitude all fully justify their claim to belong to our nation. The few exceptions are not sufficiently numerous to change my view that they were all very decent people.

The main branch of trading with which they usually start in Australia before they become sufficiently integrated into the business world to branch out in other directions is the sale of rings, chains, watches and brooches made in Rzeszow inPolish Galicia from what is called Rzeszow gold—a cheap metal amalgam, which looks and shines like gold.

One may wonder how this imitation finds so many buyers where the genuine article is so plentiful. The shape and workmanship are good and people prefer shapely ornaments even if they are not twenty-four carat. The velvet lined boxes in which these goods are sold enhance their appeal. The new arrivals in Australia, who before coming here could not even dream of a watch, a brooch, or a ring are shown the articles cleverly displayed by a Polish Jew at a cheap price and are tempted to buy it to appear in new glory at the next social occasion. The fact that the gold is imitation and that the stone is not genuine is disregarded because the shine and the fire of the gemstone is there. And so Rzeszow gold, which gives a good margin of profit, is the ladder on which the Polish Jew climbs to wealth, buying hotels on the goldfields, opening general stores or warehouses.

Quite a few Jews engaged in this Rzeszow trade lodged with Mr Silberberg. Early in the morning they used to leave with their small cases and late in the evening they would return. I became quite friendly with them and they were very helpful to me in arranging final details before departure. I found that each one of them made some sales every day and some days quite a considerable profit.

Two steamers were to leave Melbourne within a couple of days, one for London and the other for Plymouth, but all the berths were already taken, so we decided to wait for the departure of the mail packet *Marco Polo*, belonging to the Black Ball Line.

I don't understand why a company having a number of ships plying all over the world calls itself a line and not just a company, but there is the Black Ball Line, the White Star Line, and so on. The sailors have the name of the Line stitched on their navy woollen shirts and the name of the ship on which they serve on their caps. But this gear is only for fine weather and for shore leave. During stormy weather they wear coats made of oil cloth and hats with a flap covering the neck made of the same material. This gives reasonable protection when high seas pour over the ship.

We were warned that the *Marco Polo*'s captain was a very strict disciplinarian, but I was happy to hear this because it assured order on board and hence more safety. Both S. and I purchased second class tickets at £35 each, but first we took a boat to have a good look at the ship. Captain Clarke told us that it was a second class ship. Only new vessels up to twelve years old belong to first class; up to about twenty-five years they are considered second class; after that they are too old for long voyages and become coastal vessels.

Marco Polo was a three-masted clipper, built for speed. It was far from new and had already done its seventh trip to Australia. It was 1,600 tons, had decent passenger cabins and lounges, and since it had managed to come to Australia I felt sure it would manage to return under the strict Captain Clarke. The only trouble was that *Marco Polo* would not be going to sea for another three weeks.

Well, meanwhile we managed to rest after our hard years at the mines, put some weight on thanks to the excellent food at Mr Silberberg's, and visit the three Melbourne theatres. In one of these the notorious Lola Montez was dancing a newly composed dance called "The Spider".[1] The public, consisting mainly of sailors, was duly impressed.

The weather was hot and I spent a fair part of my time indoors. I had some interesting talks with a rabbi who arrived from Poland via Instanbul. He was rather condescending in his manner and as one of my Jewish friends said to me: "He is so clever that when he talks nobody can understand a thing."

The rabbi announced that he had the power to find the mother lode of gold. He was kept in the best of everything by the Jewish Melbourne community and finally announced that the location of the lode was in Bendigo.

Eventually a suitable spot was selected in Bendigo and the rabbi demanded that a tent stitched by the hand of a virgin be prepared and a young mother be placed in the middle of it on a chair. He rubbed some black shiny substance into the woman's hand, saying incantations all the time, and when the reflected sunlight from her palm struck the ground it indicated the place where to dig. Dig they did, far below the pipeclay, but did not find any gold. This I think was the demise of the rabbi's fame, because soon after someone else was elected to the position of chief rabbi.

The three weeks went very fast and the day came when we were to embark. Good old Silberberg and his daughters decided to see us off, so we went together to the river Yarra Yarra, where a small steamboat was waiting for passengers and their luggage to deliver us to the liner anchored in the bay. It took until eleven in the morning to load all the passengers, their trunks, bundles and parcels. The bell announced departure and we left the shore. We steamed up the river for nearly an hour. The Melbourne buildings got smaller and smaller until they disappeared completely.

I was leaving with mixed feelings. Resentment of the hard work and the primitive life I was forced to lead in the bush for three and a half years and the meetings with unpleasant and violent characters; but also sorrow at parting from good friends and pleasant companions. But my eyes were directed north all the time, towards my homeland where I concentrated all my hopes.

However there was little time for such reflections for we were nearing the liner which had a blue flag at the mast announcing in the seamen's code that *Marco Polo* was sailing that day. There was much activity on board. Passengers boarding and locating their cabins, small steamers, lighters and boats arriving, sailors lowering cargo down holds amidst singing and shouted instructions. Suddenly a commotion when one of the officers struck a sailor for some misdeed and the sailor dived overboard and swam towards the shore. A boat was lowered to chase the sailor, who was overpowered, brought back and locked up.

Evening arrived and tea was served, very welcome because I had had no lunch that day. At ten o'clock in the evening the mail arrived. Large leather bags filled with letters, official communications and newspapers. All the activity is at the double because midnight is only two hours away and *Marco Polo* as a mail packet is bound to leave on a specified day or it is subject to a fine.

The ship's captain undertakes to deliver the mail within a specified time and the company pays £50 fine for every day's delay. While the fine cannot be greater than what the government pays for the delivery of the mail, this is a considerable sum. Occasionally, if storm or unforeseen mishap delays a mailboat over a long period, the shipping line loses a few thousand pounds. Consequently, only those ships which do not carry mail postpone departure dates. Captain Clarke undertook to deliver the mail within sixty-seven days. He had covered the same route several times previously within this time, so we hoped he could do so again. Midnight was getting closer and *Marco Polo* hadn't moved yet, but the tug which was to tow us out was ready and I felt a slight movement and looked at my watch. It was half past eleven. Captain Clarke kept his word in spite of the becalmed seas in the bay, which prevented the use of sails.

Next morning, on 27 March at eight o'clock, we approached the dangerous narrows between the bay and the open sea. Although the place is well surveyed, it is still a hazard to passing sea traffic. We saw the wreckage of the *Schomberg*, which became a victim two weeks before we left the goldfields. *Schomberg* was a new large ship on its first journey to Australia and was also a member of the Black Ball Line. She crossed the oceans without trouble to be lost at the very goal. An officer by the name of Shaksby and a few sailors from the unfortunate vessel were returning with us to England. The Line had had big losses during the year, because apart from *Schomberg* it lost two more vessels returning from India.

We eventually passed the danger point and the tug was ready to return to port as soon as ticket control was over. Tickets are controlled to ensure that nobody stays on board who did not pay his fare. Eventually the tug left with the controllers.

Three days out a twenty-year-old stowaway was discovered below decks, but it was too late to return him. No doubt one of the sailors helped him to hide. In spite of investigation no accomplice to this smuggling aboard was discovered. The captain used him as a steward at our mess.

When we arrived in Australia the current which looked from the distance as a ribbon was on our right. This time we cut across it, an unpleasant sensation after so many years on land, because it caused pitching. We were returning by a different route. Ships which stop at South Atlantic islands and African ports to deliver their goods return by the same route which we followed arriving in Australia. However those going directly to Europe, particularly mail packets, return by a different route.

After leaving Phillip's Bay, they sail east to enter the open sea past Van Diemen's Land. To catch favourable winds and currents they sail to sixty degrees South latitude, or sail between Terra del Fuego and New Shetland, past Cape Horn, the southernmost part of America, then turn north towards the Equator. Such a journey often takes less than three months.

Notes

CHAPTER 1

1. Here the author means Geelong.
2. Phillip Tal is obscure, it may mean Phillip Valley.
3. The very high costs in Melbourne are enumerated in T.A. Coghlan, *Labour and Industry in Australia*, vol. 2, pp.796–808. He notes that in December 1852, about the time Korzelinski arrived, a two-pound loaf of bread was nine pence, eggs seven pence each and butter three shillings a pound.

CHAPTER 2

1. Jim Crow was Daylesford.
2. Here Korzelinski is exaggerating. In 1856 the total population of the Australian colonies was about 840,000, according to a calculation made in the Census of Australia for 1911 (vol.1, p.36).
3. Land laws and land prices in the colonies were very complicated and intricate. There is an analysis in Roberts, *History of Australian Land Settlement*, Chapter 8. Korzelinski is about right with his £1 an acre.
4. This is not quite right. Governor Phillip was instructed to offer approved emancipated convicts thirty acres only with twenty more for a wife and an additional ten for each child. Details are in *Historical Records of Australia*, 1 i 14.
5. This whole section on the Aboriginals should be read in conjunction with G.N. Blainey, *Triumph of the Nomads*.
6. Edward Stone Parker (1802–65), assistant protector of Aboriginals and a Methodist preacher. After travelling widely observing the Aboriginals and supporting them against Europeans, in 1841 he established the Aboriginal station of Larnebarramul (Jim Crow) at Franklinford in central Victoria. Only a few of his charges remained by 1849, and it was probably the remnant of them who were observed by Korzelinski.
7. Here the author seems to have confused the actual spear used by the Aboriginals and the spear-thrower or *wammara* (normally spelt "woomera" nowadays). Blainey in *Triumph of the Nomads* (p. 128) observes that the woomera was a wooden instrument usually about twenty-five to thirty inches long, about four inches wide, about one pound in weight and slightly scooped in shape.

8. Friedrich Wilhelm Ludwig Leichardt (1813–48?), naturalist and explorer, was born in Prussia. After some earlier expeditions, he prepared in early 1848 to explore from the Darling Downs in Queensland right across Australia to the Swan River. After leaving the Downs on 3 April 1848 he was never heard of again.
9. It is not clear what Korzelinski means by 'A fourth expedition" and exactly whom he refers to in terms of the "two brothers".

CHAPTER 3

1. For the price of food, see Ch. 1 note 3 on p. 152.
2. Forest Creek was the early name for Castlemaine.

CHAPTER 4

1. It is difficult to say who it was who first discovered gold in Australia, but Korzelinski's compatriot is a contender. Sir Paul Edmund de Strzelecki (1797–1873) went on a geological survey in 1839 and subsequently, when paying gold was discovered, started a long campaign to prove his own priority of claim.
2. Burra Burra and its copper deposits were found in 1845.
3. Gipps transmitted Strzelecki's report in a despatch of 28 September 1840, according to *Historical Records of Australia*, I xx 841, I xxi 847.
4. Edward Hammond Hargraves (1816–91), gold rush publicist, is commonly said to have discovered payable gold but this all depends on what is reckoned to be "payable" and on Hargraves's encouragement of other people.
5. Bendigo seems to have been named after an English puglist of that name, but who exactly found gold there first could not be decided when a select committee of parliament looked into the matter in 1890. Blainey in *The Rush to be Rich* (pp. 34–35) suggests that it was Margaret Kennedy, wife of the overseer of Ravenswood station.
6. A larger nugget was found at Mount Moliagul in 1869. It weighed 2,284 ounces, melted of impurities, and was sold for £9,436. It was the "Welcome Stranger" nugget, probably the largest recorded chunk of gold in the world.

CHAPTER 5

1. Adolf Knigge 1752–96, a German writer and pedagogue, author of a book entitled *Uber den Umgang mit Menchen* ("About Associating with Men").
2. Lola Montez (1818–61), dancer and courtesan, lived an erratic life and seemed able to charm many men, including the composer Liszt and the ageing King Ludwig I of Bavaria. She performed a "Spider

Dance" which was considered offensive. She opened her season in Ballarat on 16 February 1856 and publicly horsewhipped the editor of the *Ballarat Times* when he attacked her notoriety.

CHAPTER 6

1. Korzelinski here most probably means Seymour; the deep but not wide river was presumably the Goulburn.
2. This is most probably Mt Korong.

CHAPTER 7

1. The whereabouts of Gibron have not been traced.
2. Korzelinski may mean Eaglehawk.
3. This attack by bushrangers is probably the same one mentioned by Serle in *The Golden Age* (p. 82): "The most notorious bushranging exploit was the ambush of the private company's escort between Heathcote and Kyneton in July 1853. Over two thousand ounces were taken, after four of the escort of six had been shot down; only three of the large attacking party were captured and executed".
4. Lord Wellington died 14 September 1852.
5. This is a most valuable account of the Red Ribbon League and should be compared with that given by Serle, *The Golden Age*, pp. 107–18.
   The Mr Bruun mentioned by Korzelinski is possibly the "Captain" Brown mentioned by Serle: Irish-born, well-educated and an ex-Captain in the Texas Rangers.
6. This was presumably Joseph Anderson Panton (1831–1913). He was appointed assistant gold commissioner at Kangaroo Gully, near Bendigo, in 1852 and senior assistant commissioner at Bendigo the next year. He is said to have handled the diggers' protest at Bendigo in 1853 with tact and conciliation, as Korzelinski implies.
7. Serle in *The Golden Age* states (p. 107) that there were only 50,000 miners in all Victoria.
8. Charles Joseph La Trobe (1801–75), superintendent and lieutenant-governor, arrived at Port Phillip on 30 September 1839. He became lieutenant-governor of Victoria in January 1851 and was faced with the manifold problems arising from the gold discoveries. He returned to England in 1854, and died there, not in Switzerland as Korzelinski asserts later. It is not clear that he received an annual salary of £15,000.

CHAPTER 8

1. See Ch. 7 note 6 on this page.
2. *Dix livres* is French for ten pounds, while *miesa* is Polish for "of meat".

3. The truth or otherwise of this story of the origin of Jim Crow is uncertain.
4. Here Korzelinski is probably thinking of William Buckley, ' the wild white man". William Buckley (1780–1856) was transported per *Calcutta* in 1803. When that ship was at Port Phillip he ran away, the ship sailed on to Van Diemen's Land and Buckley lived with the Aboriginals until he gave himself up to a party of Europeans in 1835.

CHAPTER 9

1. See Ch. 7 note 8 on p. 154.
2. The new administrator was Sir Charles Hotham (1806–55), naval officer and governor. He was appointed to Victoria in December 1853 and ran into a good deal of trouble especially in relation to his Legislative Council and the diggers' rebellion at Eureka.
3. Korzelinski's account of Hotham's welcome should be compared with that of Serle, *The Golden Age* (pp. 156–57), with which it generally agrees.
4. Korzelinski's (second-hand) account of Eureka does not add much to existing knowledge.
5. Hotham was not made govenor-general of the Australian colonies but was elevated to full governship in February 1855.
6. Hotham died, as Korzelinski describes, on 31 December 1855, and his place was temporarily filled by Sir Edward MacArthur (1789–1872), as head of the armed forces and administrator.

CHAPTER 10

1. Korzelinski is wrong in asserting that "deportation" (transportation) was a punishment for vagrant and homeless women. Some were no doubt vagrant and without fixed abode but they were transported for criminal offences.
2. This is an interesting view of Mrs Chisholm. Caroline Chisholm (1808–77), philanthropist, has generally been regarded as virtually a saint. See Margaret Kiddle, *Caroline Chisholm* and Mary Hoban, *Fifty one Pieces of Wedding Cake*.
3. This reference has not been traced. Louisa Ann Meredith (1812–95), English poetess, travelled extensively in the colonies. She wrote three books of observations on colonial life: *Notes and Sketches of New South Wales* (1844), *My Home in Tasmania* (1850) and *Over the Straits* (1861) as well as other works.

CHAPTER 11

1. See Ch. 5 note 2 on p. 153.

Further Reading

Korzelinski's book is one of numerous accounts of life in Victoria and on the goldfields in the 1850s.

The reader who wishes to place Korzelinski in relation to other observers should consult the *Historical Studies Eureka Supplement* (Melbourne, 1965) and in particular the checklist of works relating to Eureka. Though Korzelinski was not at Ballarat, the eye-witness accounts listed in the *Eureka Supplement* cover the exact period of Korzelinski's stint on the goldfields.

Two other books should also be consulted. These are Geoffrey Serle's *The Golden Age* (Melbourne, 1963) and Geoffrey Blainey's *The Rush That Never Ended* (Melbourne, 1963). Both are necessary reading for those who wish to set Korzelinski in the bigger picture, and both have very useful bibliographies.

Korzelinski's observations on the Aboriginals should be followed up in Geoffrey Blainey's *Triumph of the Nomads* (Melbourne, 1975).

The other work used extensively in compiling the explanatory notes is the *Australian Dictionary of Biography* (Melbourne 1966).

Details of other works cited in the notes are:

Coghlan, T.A. *Labour and Industry in Australia.* 4 vols. Melbourne: 1969, first published 1918.

Historical Records of Australia, Series I. Sydney: Library Committee of the Commonwealth Parliament, 1914–25.

Hoban, Mary. *Fifty-one Pieces of Wedding Cake: a Biography of Caroline Chisholm.* Kilmore: 1973.

Kiddle, Margaret. *Caroline Chisholm.* Melbourne: 1950.

Roberts, S.H. *History of Australian Land Settlement.* Melbourne: 1968, first published 1924.

Legislative Council of N.S.W. Debates, 1851, 1852, 1853. Particularly 5 October 1953 with a statement by J. McArthur that Strzelecki and not Hargreaves discovered gold in Australia.

Colonies and East India Parliamentary Papers, Vol. 17, London: 1841.

Journal of the Royal Geographical Society 15 and 16.

Journal of the Royal Australian Historical Society (1940) part 1.

Details of Publications in which Illustrations Appear

Illustration No.	
2, 13, 20, 22	*Sketches in Victoria*, drawn on stone by S.T.G. Melbourne. Engraved by James J. Blundell & Co., Collins Street. Campbell & Ferguson, Lithographers, Melbourne. Published in parts, May 1855–1860. Ferguson no. 9922.
4, 6, 8, 11, 12, 16, 18	*Part the first. Containing 24 sketches of the Victoria gold diggings and diggers as they are*, by S.T.G. Lithographed and published by Macartney and Galbraith, Melbourne. August 1852. Issued in 2 parts.
5, 9, 14	*The gold diggers portfolio*, a series of sketches of the Australian gold fields, taken by talented artists on the spot. Engraved by Thomas Ham. Melbourne, 1854.
7	*The diggers and diggings of Victoria as they are in 1855*. Drawn on stone by S.T. Gill. Melbourne. J.J. Blundell, 1855.
10, 15, 17, 19, 21	*Five views of the gold fields of Mt. Alexander and Ballarat in the colony of Victoria*, drawn on the spot by D. Tulloch. Engraved and published by T. Ham. Melbourne, 1852.

Index